Temperate Horticulture
Current Scenario

Temperate Horticulture
Current Scenario

Dr. Daniel Jacob

Oxford Book Company
Jaipur ▪ New Delhi

Published By

Oxford Book Company

Regd. Off. "Bony Residency", Gate No. 2
Opp. Tilak Public School, Vishweswariya Nagar,
Gopalpura Road, Jaipur - 302018, India
Ph: 0141 - 2761280, Telefax: 2761381
email - oxfordbookjpr@gmail.com
website - oxfordbookcompany.com

Delhi Office: 102, 1st Floor, Satyam House,
4327/3, Ansari Road, Darya Ganj,
New Delhi - 110002, India
Ph: 011-45652440

Distributed by
KEYA INTERNATIONAL PRESS LLP
E-mail: keyainternationalpressllp@gmail.com

Published in India, 2025

ISBN : 9789380179384

Typeset by Shivangi Computers, Jaipur

Printed and bound in India by Replika Press Pvt. Ltd.

Preface

Temperate zones for horticulture cannot be defined exactly by lines of latitude or longitude but are usually regarded as including those areas where frost in winter occurs, even though rarely. Thus most parts of Europe, North America, and northern Asia are included, though some parts of the United States, such as southern California and Florida, are considered subtropical. A few parts of the north coast of the Mediterranean and the Mediterranean islands are also subtropical. In the Southern Hemisphere, practically all of New Zealand, a few parts of Australia, and the southern part of South America have temperate climates. For horticultural purposes altitude is also a factor; the lower slopes of great mountain ranges, such as the Himalayas and the Andes, are included. Thus the temperate zones are very wide and the range of plants that can be grown in them is enormous, probably greater than in either the subtropical or tropical zones.

Plants in the temperate zones benefit from a winter resting season, which clearly differentiates them from tropical plants, which tend to grow continuously. Bulbs, annuals, herbaceous perennials, and deciduous trees become more frost-resistant with the fall of sap and therefore have a better chance of passing the resting season undamaged. Another influence is the varying length of darkness and light throughout the year, so that many plants, such as chrysanthemums, have a strong photoperiodism. The chrysanthemum flowers only in short daylight periods, although artificial lighting in nurseries can produce flowers the year round. Most of the great gardens of the world have been developed in temperate zones. Particular features such as rose gardens, herbaceous borders, annual borders, woodland gardens, and rock gardens are also those of temperate-zone gardens. Nearly all depend for their success on the winter resting period.

The present book strives to highlight the current status of temperate horticulture. It provides an excellent survey of all aspects of temperate crop production and presents cutting edge research at a level which is understandable to the undergraduate population. This book will be a practical guide to practising farmers of horticultural crops and will also be a useful reference to policy makers, researchers, extension workers and students.

Dr. Daniel Jacob

Contents

1. Introduction 1

Part I: Cultivation of Temperate Fruits

2. Apple 15
3. Apricot 33
4. Cherry 45
5. Pear 59
6. Peach 73
7. Persimmon 89
8. Plum 113
9. Strawberries 127

Part II: Cultivation of Temperate Vegetables

10. Asparagus 153
11. Broccoli 165
12. Capsicum 173
13. Lettuce 187
14. Potato 201
15. Tomato 213
16. Mushroom 233
17. Turnip 259

Part III: Cultivation of Temperate Flowers

18. Chrysanthemum 269

19. Flowers and Foliage from Plants of Protea Family 279

20. Rose 291

21. Lotus 299

22. Gladiolus 305

Bibliography 317

Index 318

1

Introduction

Temperate Zone, one of the world's main climato-vegetational zones comprising a group of humid climates occupying an intermediate latitudinal position between the tropical zone equatorward and the boreal (or subarctic) climates of the polar zone poleward, and incorporating areas which have four to eight months each year with an average temperature above 10º C (50º F). The word "temperate" is, however, not wholly suited to describe all climates within this zone, for the continental variant can hardly be described as equable given its severe winters.

In geography, temperate or tepid latitudes of the globe lie between the tropics and the polar circles. The changes in these regions between summer and winter are generally mild, rather than extreme hot or cold. But in continental areas, such as central North America the variations between summer and winter can be extreme. In regions traditionally considered tropical, localities at high altitudes may have a temperate climate. The north temperate zone extends from the Tropic of Cancer to the Arctic Circle. The south temperate zone extends from the Tropic of Capricorn to the Antarctic Circle.

Within these borders there are many climate types, which are generally grouped into two categories: maritime and continental.

The maritime climate is affected by the oceans, which help to sustain somewhat stable temperatures throughout the year. In temperate zones the prevailing winds are from the west, thus the western edge of temperate continents most commonly experience this maritime climate. Such regions include Western Europe, and western North America at latitudes between 40° and 60° north.

The continental climate is usually situated inland, with warmer summers and colder winters. Heat loss and reception are aided by extensive land mass. In North America, the Rocky Mountains act as a climate barrier to the maritime air blowing from the west, creating a continental climate to the east. In Europe, the maritime climate is able to stabilise inland temperature, because the major mountain range - the Alps - is oriented east-west.

The majority of the world's human population resides in temperate zones, especially in the northern hemisphere because of the mass of land.

In the Northern hemisphere the temperate zone includes northern Europe including the middle latitudes of European Russia; north-eastern China, Russian Dongbei, Korea, and Japan; Washington State, United States, and the coastal zone of British Columbia, Canada; and the north-central and north-eastern parts of the United States and the eastern provinces of Canada. The temperate zone in the Southern hemisphere comprises much smaller land areas: the southern part of the Chilean coast; the south-eastern tip of Australia and the island of Tasmania; and the South Island, New Zealand.

TEMPERATE CLIMATE

There are two main types of climate within the temperate zone, the oceanic and the continental.

Oceanic Climate

The milder oceanic variety has an average temperature of 0º C (32º F) or above in all months of the year. This climatic zone has weather like that of the adjacent ocean, as its name implies, and is found on the western, or windward, side of mid-latitude continents. Weather and climate in the mid-latitudes are dominated by westerly airflows. Warm, moist air moving towards the poles from the tropics as a result of the general circulation of the atmosphere is deflected into a westerly airstream by the Coriolis effect of the Earth's rotation.

In the temperate mid-latitudes these tropical air masses meet cold, polar air masses moving towards the equator. These warm and cold air masses do not mix easily but instead remain clearly defined; the boundary between them is known as the polar front. The unsettled, cloudy, wet, and windy weather that characterises the oceanic temperate zone, especially in winter, is associated with the cyclones which develop at the polar front.

Cyclones, or lows or depressions, are areas of low atmospheric pressure characterised by a spiral pattern of air circulation, with winds blowing from west to east in an anticlockwise direction in the Northern hemisphere and a clockwise direction south of the equator. This circulation pattern causes the cold air masses to move eastward and equatorward, while the tropical air moves eastward and poleward in a wedge shape called the warm sector. This movement causes the polar front to buckle into two components. To the east, where the leading edge of the warm sector is replacing cold air as it moves poleward is a warm front; to the west, where polar air is moving in to replace the warm sector air, is a cold front. Because the cold front is moving faster than the warm front, the warm sector is gradually squeezed; eventually the cold air masses on either side of the warm sector meet and the cyclone decays. The warm and cold fronts are associated with cloud and precipitation, usually heavier at the cold front. There are also often sharp changes of temperature as successive fronts pass over an area.

Although cyclones are a key feature of the oceanic temperate climate generally, there are differences between the Northern and Southern hemispheres. In the former, the presence of mountain ranges and the relative distribution of land and sea means that cyclones have preferred areas of formation: notably the so-called Aleutian low in the North Pacific and the Icelandic low in the North Atlantic. There is also a seasonal difference in the Northern hemisphere: in the winter the belt of highest cyclone frequency is centred around 50° north; in the summer it moves poleward to 60° north and the area is more strongly influenced by south-westerly air streams of tropical maritime origin. In the Southern hemisphere the lack of significant land masses in the temperate zone means that the cyclone belt is almost continuous around the globe; it is also unaffected by seasonal changes, remaining between 50° south and 60° south throughout the year.

The other major weather system found in the temperate zone is the anticyclone or high. Anticyclones are persistent regions of high pressure characterised by low cloud, absence of precipitation, and low wind speed-although winds may be stronger on the edge of the high. Because of their size and stability, anticyclones tend to block the passage of depressions. Anticyclones draw polar, continental air masses towards the western margins of continents. These air masses are dry at any season; in winter they are very cold, in summer they are very warm. Anticyclones thus provide protracted periods of stable weather, in contrast to the changeable cyclonic weather which tends to prevail in the oceanic temperate zone.

The description "temperate" is relatively appropriate for the oceanic climatic type. The mean (average) air temperature in summer is below average for the latitude, while the winter mean is above average. Neither the diurnal nor the annual range in temperature is extreme in areas with oceanic climates, such as the British Isles. The weather is comfortable during the summer, although not ideal for ripening cereal crops. For example, Oxford, in southern England, has a mean air temperature in summer of 15.7º C (60.5º F), with a daily mean maximum of 20.8º C (69.5º F) and a daily mean minimum of 11.4º C (52.5º F), a relatively small range with neither hot days nor cold nights. Winters are similarly mild with a mean air temperature of 4.0º C (39º F), a mean maximum of 7.1º C (44.6º F), and a mean minimum of 1.5º C (34.7º F).

Precipitation in the oceanic temperature zone varies strongly with altitude and usually reaches a maximum in late autumn-early winter. In Great Britain, west coast locations typically receive about 1,000 mm (39.4 in) a year, but this may increase to more than 3,000 mm (118 in) in mountainous locations, such as the Lake District in north-western England, or Snowdonia in Wales. East of mountainous areas, in the rain shadow, totals fall.

As distance from the sea increases, the rainfall regime becomes more continental, with a summer as well as a late-autumn peak of precipitation. Snowfall increases not only with altitude but also with distance north. However, the number of frosts is usually determined by highly localised factors and can vary greatly within relatively small areas.

The temperate oceanic climate is one of the cloudiest on Earth. The Sun is normally seen for only 30 to 40 per cent of the total possible time.

Continental Climate

Although scarcely "temperate", at least during the winter months, the temperate continental climate is found in some of the Earth's most populous and economically developed areas such as the American Midwest and eastern Europe. The continental temperature regime is much more strongly seasonal than that of the oceanic type, combining extremely cold winters, including prolonged snow cover, with hot summers. The large annual temperature range is quite clearly a product of continentality-that is, of the strong heating of the land in summer and of the equally strong cooling in winter. The continental climate is confined to the Northern hemisphere because there are no large land masses in the Southern hemisphere at suitable latitudes.

In general, the continental climate becomes more severe from south to north and from the east coast inland, especially in winter. Bucharest, Romania, has a mean summer temperature of 21.5º C (70.7º F) and a mean winter temperature of -2.0º C (28.4º F). Moscow, Russia, has mean summer and winter temperatures of 17.6º C (63.5º F) and -9.6º C (14.5º F) respectively. The combination of long, hot summers and a frost-free period of 150-200 days each year makes these regions highly productive agriculturally. At Urbana, Illinois, in the heart of the American corn (maize) belt, the average July temperature is 23.9º C (75º F). The mean maximum for July is 30º C (86º F) and, because of slow cooling at night, the mean minimum is 17.8º C (64º F). Only occasional incursions of Arctic air interrupt the succession of hot, often humid, days. Severe winters are equally notable. For example, Dubuque, Iowa, United States, has 133 days with air frosts and 56 days when the temperature fails to exceed freezing.

Most places in the temperate continental climatic zone receive less than 750 mm (29.5 in) of precipitation annually, with a summer maximum. Strong convection commonly creates dramatic thunderstorms during the summer.

VEGETATION AND SOILS

The natural vegetation of the temperate zone is largely determined by local climate. On the equatorial side of the tundra region of the polar zone, temperate boreal forest, comprising evergreen conifers, is found. Various species of spruce are the dominant trees, but pines, larches, and firs are also common. Near the tundra, dwarf birch becomes important. Despite its poleward location, the boreal forest is quite productive, and the region is of great importance for timber production. Because of the low temperatures, organic matter tends to decay rather slowly in the boreal forest. A deep layer of leaf litter may build up on the forest floor, which can fuel forest fires-a common hazard. The litter also produces organic acids which percolate through the underling soil, transporting iron downward and producing the distinctive podzol soil type. These are generally infertile soils, having been leached of most of their nutrients by the acid-rich ground water.

Deciduous temperate forests lie equatorward of the boreal forests. Here, broad-leaved trees lose their leaves in autumn and lie dormant through the winter. The dominant trees are species of oak, elm, beech, chestnut, and maple. Lesser acidity and greater activity by bacteria and other soil fauna lead to the formation of fertile brown-earth soils. Extensive areas of grassland are found in the interiors of continents, where precipitation amounts are too small to support tree growth-usually less than 500 mm (19.7 in). Various names are given to these grasslands, including prairie, steppe, and pampas. There is some debate among scientists as to whether the grassland is a natural climax vegetation-that is, the mature, stable, natural vegetation that occurs at the end of the process of environmental succession-or whether it is the product of human management over many centuries.

The famous chernozem or black-earths are particularly associated with the grassland. They are highly productive soils that support intensive arable agriculture in areas like the United States, Canada, Russia, and Ukraine. Another important soil type, found over large areas of the temperate zone, is loess. Formed from silt-sized particles of sedimentary material created by glacial action, mainly during the Pleistocene ice age, and deposited by wind action, these light soils are easily worked and very fertile. However, they can be easily eroded ʼn farmed carelessly.

IMPACT OF GLACIATION

The temperate zone has been greatly affected by the climatic changes that occurred during the Pleistocene epoch of the Quaternary period. There have been numerous ice ages in the Earth's history, most recently during the Pleistocene, when ice sheets expanded and contracted producing glacial and interglacial periods respectively. There have been at least eight such cycles in the past 2 million years, each lasting on average

125,000 years. Only 10 per cent of each cycle was as warm as the present day. The last glacial maximum was about 18,000 years ago. Huge amounts of water were stored in the ice sheets which developed in North America and northern Europe, and sea levels changed, falling dramatically. Temperatures were then about 5 to 7º C (9 to 12.6º F) cooler.

Many temperate landscapes bear the legacy of formerly glacial or periglacial conditions. In the British Isles, for example, the maximum extent of glaciation occurred during the penultimate glacial period, about 125,000 years ago, when ice sheets extended as far as South Wales and the Thames valley. In mountainous regions such as the Lake District and the Cairngorms, evidence of former glaciation is found in features such as cirques, U-shaped and hanging valleys, and roche moutonées. South of the glacial limit, there was a severe periglacial climate, equivalent to the tundra regions of the Arctic today. Evidence of permafrost activity is found on the chalk downs of southern England and in much of France, for example.

HUMAN IMPACT

The impact of human beings on the vegetation and landscapes of the temperate zone, especially in the Northern hemisphere, has been profound. Hundreds-and in many areas, thousands-of years of intensive human settlement and economic activity have left very little of the native vegetation of the zone undisturbed. Even apparently wild landscapes such as the moorlands of the Scottish Highlands tend to be a result of human interference.

The development of agriculture in the temperate zone was closely linked to the felling of the forests which covered much of the region, providing access to the fertile forest soils. Although deforestation of the temperate zone began in prehistoric times it progressed at different rates in different areas, reflecting in large part variations in population pressures. Most of England's woodlands, for example, had been felled by the mid-14th century to provide both agricultural land and fuel. In China clearance of most of the temperate forest was completed even earlier; the resulting shortage of fuel was an important reason for the development of the energy-efficient stir-fry techniques which characterise Chinese cookery. In contrast, the main deforestation of continental western Europe and North America did not take place until the 18th and 19th centuries-when the population expansion associated with the Industrial Revolution and, in the case of North America, large-scale immigration necessitated land clearance to meet the food, fuel, and construction needs of the rapidly expanding cities.

As a result of forest conservation and management the total area under temperate forest has stabilised, and even increased, during the 20th century. The forested area in Britain has doubled since 1900 to 10 per cent, while in the United States there has been a significant expansion in the area of east-coast forests, partly as a result of the abandonment since 1940 of many small farms in the face of agricultural modernisation. However, very little of this is the original native forest: in western Europe and the United States only about 1 per cent of the original native forest is left, the rest is re-growth or, increasingly, commercial timber plantations.

Because such plantations normally have only one or two species of trees-often non-native species-they provide a much more limited habitat than native forests. As a result, many temperate-zone plants and animals have become rare, or are at risk. One notable early casualty of human expansion into temperate forest areas was the passenger pigeon. This bred in colonies of many millions in the deciduous forests of eastern North America until the 1800s. Subsequently, loss of habitat and aggressive hunting led to its rapid decline. The species had become extinct in the wild by 1894; it became totally extinct in 1914, when the last captive-bred bird died.

The grasslands of the North American prairies and the Eurasian steppe were equally affected by the rapid expansion of the human population of the temperate zone during the 18th and 19th centuries, and the consequent need to expand agricultural production. Russians began to colonise the steppe during the 16th century, but the ploughing up of the Eurasian grasslands on a large scale only began in the 18th century and reached a peak under Soviet rule, when large-scale mechanisation led to most of the steppe being brought under cultivation. Significant cultivation of the North American prairies began about a century later.

In the subsequent 150 years about two thirds of the native vegetation of the Canadian prairies and about half that of the US prairies has disappeared as the land has been converted into huge cereal farms and ranches. The resulting loss of habitat, combined with hunting, has led to the extinction, or near extinction, of most of the large mammals which once roamed the temperate-zone grasslands, including the bison, the tarpan, the pronghorn, and the saiga antelope. The bird life of the plains has also been greatly affected by human activities; many have become rare or are on the endangered list, including the steppe eagle, great and little bustard, the demoiselle crane, and the prairie chicken.

The intensive cultivation of the grasslands has led to other problems. In particular, over-grazing and ploughing led to the loss of the vegetation cover which protected the soil from being blown away by the wind-particularly during periods of drought. As a result, both the prairie and steppe have experienced major problems of soil degradation and erosion over the past 70 years, leading in the United States to the creation of what has become known as the Dust Bowl.

Pollution is the other major problem which faces the temperate zone. In the 1950s there was great concern about smoke and sulphur-dioxide pollution of the atmosphere from coal-burning. Clean-air legislation and changing sources of domestic heating have largely eradicated this problem in urban areas. However, long-distance transport of sulphurous pollutants has caused air-pollution problems. Although often referred to as acid rain, such pollution also includes the dry deposition of acid particles and gases as well. Acid deposition threatens trees and fish in vulnerable locations such as the granite mountains of southern Scandinavia and Scotland, and the Black Forest of Germany. Today, a good deal of such pollution comes from large coal-burning power stations and from nitrogen oxides emitted from car exhausts.

Perhaps a more pressing problem of atmospheric pollution in the temperate zone concerns the emission of so-called greenhouse gases such as carbon dioxide and chlorofluorocarbons (CFCs). Many scientists now believe that these gases have caused global warming over the last century and will continue to do so for the foreseeable future. Much research is being undertaken to model global climate change in order to understand better the changing weather patterns and other impacts which may result.

Water pollution is another problem caused by the urbanisation and intensive economic development of the temperate zone. Streams and rivers in many areas are contaminated by industrial waste products and the discharge of large amounts of partly treated or untreated sewage. Modern, intensive agricultural practices have added to the problem. The run-off of chemical fertilisers, through contaminated surface and groundwater, into rivers and lakes can lead, for example, to eutrophication and a resulting loss of oxygen for fish and other aquatic creatures.

TEMPERATE BIODIVERSITY

Temperate climates are those without extremes of temperature and precipitation. The changes between summer and winter are generally invigorating without being frustratingly extreme. There are two types of temperate climate: maritime and continental. The maritime climate is strongly influenced by the oceans, which maintain fairly steady temperatures across the seasons. Since the prevailing winds are westerly in the temperate zones, the western edge of continents in these areas experience most commonly the maritime climate. Such regions include Western Europe, in particular the UK, and western North America at latitudes between 40 and 60° north.

Continentality increases inland, with warmer summers and colder winters as the effect of land on heat receipt and loss increases. This is particularly true in North America, where the north-south aligned Rocky Mountains act as a climate barrier to the mild maritime air blowing from the west. Maritime climate, on the other hand, penetrates further into Europe where the major mountain range - the Alps - is orientated east-west.

Temperate forests are often called deciduous forests. In a temperate forest, most of the trees lose their leaves in the winter. During the fall, when the weather gets cooler, the trees begin to shut down. Their leaves turn beautiful shades of colours. Come winter time, the leaves fall off of the trees. Why do you think this is so?

The deciduous trees must lose their leaves, because water is not available for the leaves to survive. Also, the trees are not strong enough to keep their leaves and hold up all the snow that comes in the winter. Here are some interesting temperate forest facts:

— The largest tree in the world is found right here in the United States. You can find it in California. This giant sequoia tree is 275 feet tall and 95 feet around. Boy, that's HUGE!

— The largest forest in the world covers parts of Scandinavia and northern Russia. It has over 3.5 million square miles of land covered in trees!

— Most hardwood trees are used for firewood, construction, or art. Lots of forests are being cut down for farm land.

In a temperate forest, there are four definate seasons. In some biomes, like the tundra, it is usually cold, but not in the temperate forest. Spring time brings new life to the trees and plants, warm temperatures, and rain which helps everything grow. Summer is hot, and everything is green for all the trees now have leaves. Fall is cool. The trees are beginning to change colours such as red, yellow, and bright orange. Winter months are extremely cold. All the trees loose their leaves.

The average temperature for a year in the temperate forest is about 50 degrees F.

The forests usually gets about 30 to 60 inches of every year. Look at the chart below to compare the rainfall of thetemperate forests with the other biomes. The temperate forests are always changing due to the weather, animals, and people.

TEMPERATE FORESTS

A forest has many layers of plant life. There are the small plants such as shrubs, moss, ferns, and lichens which grow quietly beneath the shadows of the tall trees.

The trees found in a temperate forest are called hardwoods. This means the trees loose their leaves in the winter. Their trunks are also made of a bark that is very hard.

Most of the trees are maple, birch, beech, oak, hickory, and sweet gum. There are many other types of trees found in the forest too.

A temperate forest also contains a few pine trees and other coniferous trees. The Taiga is usually found north of the temperate forests, but sometimes they overlap.

Animals in a forest come in all different shapes and sizes. Below are a few examples of the kinds of animals you might find in a forest. Have you ever seen any of these animals before?

Cardinal - A cardinal is a bird that is often found in the United States. You can tell it is a cardinal by its bright red feathers.

Squirrel - Here is an animal that you might have seen before. A squirrel is a furry animal with a big, bushy tail. It eats nuts and seeds and lives in the trees.

Bear - There are many types of bears around the world. There are grizzly bears, black bears, and polar bears just to name a few. Take a look at the site called The Cub Den to learn more about bears.

TEMPERATE HORTICULTURE

Temperate regions have been endowed with a wide range of agro-climatic conditions due to which a large number of horticulture commodities like fruit crops, flowers, vegetables, mushrooms, hops, tea, medicinal and aromatic plants etc. are successfully grown here. Amongst the fruit crops, perhaps all kind of fruits grown in the country, except those which are grown in the warm humid coastal regions, can be produced in the region.

In addition to fruits, vegetable and flower cultivation is undertaken for off season supplies to the plains, while the mushroom cultivation is being taken up even by the landless for the augmentation of their income. Beekeeping is a necessary activity adjunct to the horticulture industry, which besides providing pollinating agents for improvement of fruit set and productivity, also results in the production of economic apiculture products like honey and bee wax. The horticulture industry has, therefore plays a great role in the amelioration of the rural economy of the State with respect to:

(i) generation of sources for cash income to the rural people,

(ii) generation of employment opportunities in the pre and post harvest sectors of the horticulture industry,

(iii) provision of nutritive foods in the form of fruits, vegetables, nuts, mushrooms, honey, etc.,

(iv) satisfaction of the aesthetic needs of the people, and

(v) development of a sustainable system of permanent agriculture in the hilly areas.

Changing Trends

If we analyse this changing scene in agriculture, we can notice that the traditional agriculture. Which was a 'way of life' for our farmers is now becoming a 'business proposition'. In the traditional farming there was no much change in the cropping pattern, cultivation practices etc. It was based on the experiences transmitted from father to the son. However, with the developments taking place due to five-year plans and technologies developments in agriculture, traditional farming is changing into modern farming. Traditional farming is slowly becoming absolute and uneconomic. Traditional farming was more or less self-sufficient. No, farming is becoming market oriented. The needs of the farmers are increasing. He has to purchase many things such as high yielding seeds, fertilisers, pesticides, machinery etc. from the market. As a result, his investment and financial needs are increasing. Naturally, he has to produce and get income to meet the costs and also to make some profit. This, the costs, returns, markets, profits of the enterprise become significantly important. This is nothing but 'Agri-economics'.

With increasing population, rapid urbanisation and growing export markets, the demand for farm products is increasing and is likely to increase in the near future. However, the competition is also likely to increase. The consideration of economic aspects in the production process is inevitable.

As indicated above, there has been a technological break-through in agriculture in recent years. New non-traditional crops, new varieties of crops, new methods of cultivation are coming in very fast and farmers are adopting the same. A large number of farm products are being produced for exports. However there are specifications about the size, colour, quantity, taste, packaging etc. Which farmers should know. Farmer has to consider all these aspects and consider the costs and returns before entering into the venture.

Farmer has several enterprises on the farm. He has to consider the economics of each enterprise separately also of the farm as a whole. It helps in decision making and proper planning of the farm. Now, the time has come that every activity on the farm has to be viewed from the perspective of economics.

Along with the adoption of new technology in farming, the problems faced by the farmer's fare also increasing. There are problems of soil and water management, choice of crops, technical know-how, pests and diseases, natural hazards, marketing, finance, surplus production, price fluctuations and so on. In finding the solutions for these problems, economic criteria are to be applied.

India is a vast country with varied climate, soils, and ecological conditions. In addition to this, individual farmer is having his own set up of resources and socio-economic situation. In solving the problems of individual farmers all these situational factors are to be taken into account.

An important ray of hope, which one can notice in this complex changing scenario of agriculture, is that a new generation of farmers who are more educated, young and energetic have taken up to this enterprise. In addition, many non-farming community entrepreneurs are also attracted towards agriculture. They are very keen on getting more knowledge about the new technology. Many of them are innovative and experimenting of their own. Naturally, they are more economics oriented.

REFERENCES

Adams, C K, 2008. *Principles of Horticulture*. Butterworth-Heinemann.

Budelman, A. & Defoer, T. 2000. *Managing soil fertility in the tropics - PLAR and resource flow analysis in practice. Case studies from Benin, Ethiopia, Kenya, Mali and Tanzania.* Amsterdam, The Netherlands, Royal Tropical Institute.

Hawthorne, 2003. *Complete book of plant propagation,* Reed Books.

McGee, J.R. and Kruse, M. 1986. "Swidden horticulture among the Lacandon Maya". University of California, Berkeley: Extension Media Center.

Part I : Cultivation of Temperate Fruits

REFERENCES

Billy, Simon. *Horticulture Farm Manager*. The Kerr Center for Sustainable Agriculture, Poteau, OK. Personal communication. May 1997.

Bordelon, B. 2002. *Grape Cultivars for the Upper Midwest*. Purdue University, Department of Horticulture and Landscape Architecture.

Chan, H.T. Jr. 1984. (Ed.) *Handbook of Tropical Foods*. New York/Basel: Marcel Dekker.

Chandler, William Henry. 1950. *Evergreen Orchards*. Philadelphia: Lea and Febiger.

2

Apple

Apple trees grow in the temperate regions of the world, generally in the latitudes between 30° to 60° north and south. Apple trees are best adapted to places where the average winter temperature is near freezing for at least two months, though many varieties can withstand winter temperatures as low as -40°C (-40°F).

Apple trees are deciduous, which means their leaves fall off at the end of the growing season. Apple trees belong to the Rosaceae family, and in the spring you can see the beautiful white flowers on an apple tree looking like tiny roses. Taxonomists have called the cultivated species of apple tree *Malus domestica,* and the wild species *Malus pumila.* Apples and pears are "pomes" because they have a paper-like core, a fleshy layer around the core, and an outer skin. The volume of a ripe apple is about 20% air (that's why they float), and 80% water.

Apple (Malus pumila) is commercially the most important temperate fruit and is fourth among the most widely produced fruits in the world after banana, orange and grape. China is the largest apple producing country in the world.

The apple is the pomaceous fruit of the apple tree, species *Malus domestica* in the rose family Rosaceae. It is one of the most widely cultivated tree fruits. The tree is small and deciduous, reaching 3 to 12 metres (9.8 to 39 ft) tall, with a broad, often densely twiggy crown. The leaves are alternately arranged simple ovals 5 to 12 cm long and 3-6 centimetres (1.2-2.4 in) broad on a 2 to 5 centimetres (0.79 to 2.0 in) petiole with an acute tip, serrated margin and a slightly downy underside. Blossoms are produced in spring simultaneously with the budding of the leaves. The flowers are white with a pink tinge that gradually fades, five petaled, and 2.5 to 3.5 centimetres (0.98 to 1.4 in) in diameter. The fruit matures in autumn, and is typically 5 to 9 centimetres (2.0 to 3.5 in) diameter. The center of the fruit contains five carpels arranged in a five-point star, each carpel containing one to three seeds.

The tree originated from Central Asia, where its wild ancestor is still found today. There are more than 7,500 known cultivars of apples resulting in range of desired characteristics. Cultivars vary in their yield and the ultimate size of the tree, even when grown on the same rootstock.

Figure 1. A branch of apple tree

At least 55 million tonnes of apples were grown worldwide in 2005, with a value of about $10 billion. China produced about 35% of this total. The United States is the second leading producer, with more than 7.5% the world production. Turkey, France, Italy, and Iran are also among the leading apple exporters.

Apples originated in the Middle East more than 4000 years ago. Spreading across Europe to France, the fruit arrived in England at around the time of the Norman conquest in 1066. The center of diversity of the genus *Malus* is the eastern Turkey. The apple tree was perhaps the earliest tree to be cultivated, and its fruits have been improved through selection over thousands of years. Alexander the Great is credited with finding dwarfed apples in Asia Minor in 300 BCE; those he brought back to Macedonia might have been the progenitors of dwarfing root stocks. Winter apples, picked in late autumn and stored just above freezing, have been an important food in Asia and Europe for millennia, as well as in Argentina and in the United States since the arrival of Europeans. Apples were brought to North America with colonists in the 1600s, and the first apple orchard on the North American continent was said to be near Boston in 1625. In the 1900s, irrigation projects in Washington state began and allowed the development of the multi-billion dollar fruit industry, of which the apple is the leading species.

PRODUCTION TRENDS

The area under apple cultivation in India increased by 24% from 1.95 lakh ha. in 1991-92 to 2.42 lakh ha. in 2001-02 although production increased by less than 1% (i.e. from 11 to 12 lakh tones). It is mostly grown in the states of Jammu & Kashmir, Himachal Pradesh, Uttaranchal, Arunachal Pradesh and Nagaland.

J&K has remained the leading apple producer accounting for 60% of the total production in the country. In HP also, apple is the most important crop accounting for about 90% of the total horticultural production. These two States accounted for 76% of the area and 94% of the production of the crop in the country in 2001-02.

Apples are mostly consumed fresh but a small part of the production is processed in to juices, jellies, canned slices and other items. India's share in the total world apple production is merely 2.05 per cent. Only around 1.6% of the country's production gets exported.

An agri-export zone has been established in Himachal Pradesh covering the districts of Shimla, Siramour, Kullu, Mandi, Chamba and Kinnaur considering the vast potential for increasing exports. Initial targets are for exporting apples to neighbouring countries as well as to west Asia and to the south-east Asian countries. With increase in production of apples in the State, the State government is taking effective steps for its marketing besides providing packing material, transportation, ensuring adequate procurement of apples under Market Intervention Scheme, so that the growers get remunerative prices in the markets. Arunachal Pradesh is one of the ideal locations for apple cultivation. If high yielding varieties of apple are introduced in the state it can make a big headway by exporting the produce to Bangladesh which presently depends on Bhutan besides meeting the internal demand of the northeastern markets. Efforts are being made by the NHB and the State Govts. for imparting modern technologies to the growers in apple producing states.

CULTIVATION

Like most perennial fruits, apples ordinarily propagate asexually by grafting. Seedling apples are an example of "Extreme heterozygotes", in that rather than inheriting DNA from their parents to create a new apple with those characteristics, they are instead different from their parents, sometimes radically. Most new apple cultivars originate as seedlings, which either arise by chance or are bred by deliberately crossing cultivars with promising characteristics. The words 'seedling', 'pippin', and 'kernel' in the name of an apple cultivar suggest that it originated as a seedling. Apples can also form bud sports (mutations on a single branch). Some bud sports turn out to be improved strains of the parent cultivar. Some differ sufficiently from the parent tree to be considered new cultivars.

Breeders can produce more rigid apples through crossing. For example, the Excelsior Experiment Station of the University of Minnesota has, since the 1930s, introduced a steady progression of important hardy apples that are widely grown, both commercially and by backyard orchardists, throughout Minnesota and Wisconsin. Its most important introductions have included 'Haralson', 'Wealthy', 'Honeygold', and 'Honeycrisp'.

Apples have been acclimatised in Ecuador at very high altitudes, where they provide crops twice per year because of constant temperate conditions in a whole year.

Agro-climatic Requirements

Apple can be grown at altitudes 1,500-2,700 m. above m.s.l. in the Himalayan range which experience 1,000-1,500 hours of chilling. The temperature during the growing season is around 21-24^0 C. For optimum growth and fruiting, apple trees need 100-125 cm. of annual rainfall, evenly distributed during the growing season. Excessive rains and fog near the fruit maturity period result in poor fruit quality with improper colour development and fungal spots on its surface. Areas exposed to high velocity of winds are not desirable for apple cultivation. Loamy soils, rich in organic matter with pH 5.5 to 6.5 and having proper drainage and aeration are suitable for cultivation.

Tree Purchase and Preparation

Purchase a healthy 1-year-old nursery tree, 4 to 6 feet tall, with a good root system. A small tree with a good root system will transplant better than a large tree. When you get the tree, protect it from injury, drying out, mouse or vole damage, freezing, and overheating. If the roots have dried somewhat, soak them in water for about 24 hours before planting. If you are unable to plant the tree immediately, there are two options:

1) Wrap the roots in plastic along with moist sawdust or newspaper, and place the tree in a refrigerator or cooler at 40 F. Never store the tree with fruit or vegetables, as ethylene gas from ripening foods can kill young trees.

2) "Heel-in" the tree. To heel-in a tree, dig a trench and place the tree roots evenly in it, cover the roots with soil, sawdust or peat, and water the tree thoroughly. The tree can be kept for several weeks using this method before permanently planting.

Planting the Tree

In North Carolina, trees can be set from late fall to early spring. To plant the apple tree, first dig a hole approximately twice the diameter of the root system and 2 feet deep. Place some of the loose soil back into the hole and loosen the soil on the walls of the planting hole so the roots can easily penetrate the soil. Spread the tree roots on the loose soil, making sure they are not twisted or crowded in the hole. Continue to replace soil around the roots. As you begin to cover the roots, firm the soil to be sure it surrounds the roots and to remove air pockets. Do not add fertiliser at planting time as the roots can be "burned". Fill the remainder of the hole with the loose soil, and press the soil down well. Important: The graft union must be at least 2 inches above the soil line so that roots do not emerge from the scion. When you have finished planting the tree, water well to eliminate air pockets and provide good contact between the roots and the soil.

Training and Pruning

Central Leader Trees

Proper training and pruning of fruit trees is essential to the development of a strong tree framework that will support fruit production. Properly shaped trees will yield high-quality fruit much sooner and will live significantly longer. Regular pruning and training will also maximise light penetration to the developing flower buds and fruit. Additionally, a well-shaped tree canopy permits adequate air movement through the tree, which promotes rapid drying to minimise pest problems.

A central leader tree has one main, upright trunk, called the "leader". Branching should begin on the leader 24 to 36 inches above the soil surface to allow work under the tree. The first year, 3 to 4 branches, collectively called a "scaffold whorl", are selected. The selected branches should be spaced uniformly around the trunk, not directly across from or above one another. The major lateral branches are commonly referred to as scaffold branches on which the central leader tree is "built". Above the first scaffold whorl should be an area of 18 to 24 inches, called a "light slot", without any branches to allow light to reach all lower leaves and fruit. This light slot is followed by another whorl of scaffolds. Maintain alternating scaffold whorls and light slots up the leader to the desired maximum tree height. The shape of a properly trained central leader tree is like that of a Christmas tree.

Newly Planted Trees

After early winter planting, wait until just before the buds start to grow in the spring to "head", or cut, the unbranched central leader to 36 inches above the soil surface to encourage new lateral branching. When new growth is 3 to 4 inches long, identify the most upright shoot that will continue to be the central leader. Leave it and remove all new shoots growing 3 to 4 inches immediately below this new terminal to prevent competition.

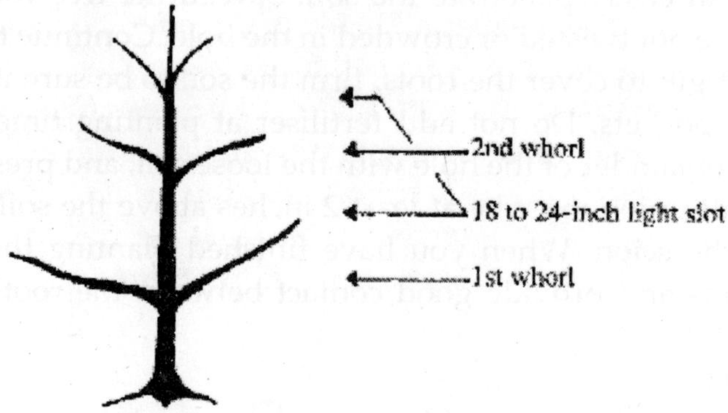

Figure 2. Side view of a central leader tree

This will also encourage lateral growth in the area 6 to 14 inches below the cut tip of the young tree. Branches that form 6 to 14 inches below the cut tip of the tree are less vigorous, less upright, and easier to train as productive scaffold limbs. When the lateral branches, or scaffold branches, are 3 to 6 inches long, they should be spread to a wider crotch angle that will provide a stronger framework for fruit production. Toothpicks or clothespins can be used to prop the young branches out to a 50 to 60 degree angle. This angle will slow vegetative growth, promote lateral branches, and allow the tree to initiate flowers and produce fruit sooner.

Scaffold Training

Improperly trained fruit trees have very upright branch angles, which result in excessive vigor and serious limb breakage under a heavy fruit load. Larger branches can be spread out using short wooden boards with a notch cut in each end for the branch to fit into. Hanging weights on the branch or tying it down with string wrapped loosely around the limb are other methods for spreading the branches. All upright growth from scaffold branches should be either pulled down to a horizontal position or removed when it is 3 to 4 inches long.

Dormant Pruning vs. Summer Pruning

Pruning the tree during the winter, while it is dormant, will invigorate the tree and cause it to grow and branch more the following season. To promote scaffold branch development, cut the central leader 20 to 28 inches above the highest usable scaffold whorl during the dormant season. It is best to do dormant pruning in late winter or early spring, after the risk of severe freeze is over. Be sure to remove any dead or diseased wood and dried apples at this time as well. After the tree resumes growth in the spring, continue to train the scaffold branches of the tree as you did the previous growing season. Select a new upright shoot to continue the central leader, and remove all new shoots 4 inches below it. Also select the branches to form another whorl of 4 to 5 scaffold branches. Prop all lateral branches out to a 50 to 60 degree angle. Summer pruning will devigorate the tree and cause it to grow less in that growing season. Remove all undesirable branches directly across from one another on the central leader when they are 3 to 4 inches long. Also, select lateral branches that are spaced uniformly around the leader to prevent crowding as the limbs grow in diameter.

Fruit Thinning

Apple trees often set a heavier crop of fruit than the limbs can withstand. To ensure good fruit size, return bloom for the following year, and to prevent tree breakage, it is necessary to thin the fruit. Every apple blossom results in a bloom cluster of 5 to 6 blossoms. Apples should be thinned when they are about the size of a dime. Cut off enough fruit so that the remaining apples are spaced 4 to 6 inches apart, and leave only one fruit per cluster. It may seem like very few fruit remain, but you will harvest higher-quality fruit, potentially reduce insect and disease problems, and increase the chances for a full crop the next season.

Slender, Spindle-Type Tree Training and Pruning

With a slender, spindle-type training system, several components must be integrated to ensure success. First, the tree must have a permanent support system the height of the entire tree to promote rapid tree growth and support the tree with heavy crop loads. A common support is a 10-ft-long piece of conduit or galvanised pipe 1 inch in diameter which is placed approximately 6 inches away from each tree and driven 2 ft into the ground. A 2- to 3-inch diameter treated w[illegible]n post works as well. Trellis systems can also be used successfully with high tensile wire. A typical trellis system will have smaller posts at each tree (i.e. bamboo stake or ½-inch conduit) attached to a top wire approximately 8 ft above the ground with large anchored posts every 50 to 100 ft to stabilise the top wire. Trellis systems are also used entirely with the wires spaced approximately every 2 ft apart to a height of 8 ft. Independent of the type of support system selected, the trees must be loosely tied to this support system every 15 to 24 inches along the entire leader. Second, the trees must receive adequate, uniform moisture during the growing season to maintain tree growth, and irrigation is highly recommended as an insurance against drought. Third, all weed competition in the tree row should be eliminated with herbicides to at least the dripline on each side of the tree.

The slender, spindle-type tree is a modified miniature central leader tree with many more limbs per tree. The trunk or central leader is maintained continuously to a maximum height of 8 to 10 ft. The branching should begin at approximately 24 to 32 inches above the soil surface. The slender, spindle-type tree discussed in this leaflet has two permanent whorls of scaffolds at maturity, spaced approximately 10 to 12 inches apart to allow for adequate light penetration and optimal fruiting in the lower portion of the tree. During the formative years all lateral branches are left on the tree to optimise early production, branches are thinned out to the two permanent whorls as shading becomes a problem in the 3rd to 5th year or later as required. Above the second whorl of branches, all branches are temporary and are renewed every 2 to 4 years. The maximum branch spread of this tree outward from the leader is approximately 3 to 5 ft.

A misconception exists that "dwarf" trees only grow 8 ft tall and only spread 3 to 4 ft, when in fact they will grow 10 to 15 ft tall and spread 6 to 8 ft outward from the tree if not properly managed. To train a slender, spindle-type tree to an allotted space and ensure optimal productivity, it is necessary to use techniques that will maintain leader growth as well as encourage lateral branching. If a tree has a maximum height of 10 ft and there is 2 ft of the tree height without branches, then the tree has lost 20% of its potential productivity. In order to avoid this problem, training techniques for a slender, spindle-type tree, from planting to maximum productivity, are discussed below.

Leader Management

After planting an unbranched "whip," the traditional leader-management technique would be to head the leader or cut off the leader approximately 30 to 34 inches above ground just before budbreak. After approximately 3 to -4 inches of new growth, a new shoot would be selected as the leader from the 3 or 4 most terminal upright shoots and then all shoots beneath the new leader for approximately 4 inches would be removed. If a well-branched tree is planted with 5 to 9 usable branches with wide crotch angles, starting approximately 24 inches above the soil, then the trees should produce fruit at least 1 year earlier than a whip. The traditional leader-management technique with a feathered tree would be a modified heading of the leader at 10 inches above the highest usable branch, with usable branches being those that have lower branches within 6 inches. Frequently what is observed in North Carolina is that there are lower branches and then a 1- to 2-ft length of the leader without branches and then several more branches. In this case, the leader would have been cut off during the dormant season

10 inches above the lower branches to ensure continuous branching along the leader. This is not a preferred technique, as the pruning would invigorate tree growth and delay fruit production. Techniques to be discussed in the remainder of this leaflet are those that may be used to produce a high density orchard system which has the potential for early fruit production. The best leader-management techniques are those that do not rely upon pruning to maintain the tree shape. Research has shown that any pruning of young trees will reduce or delay fruit production early in the life of the orchard.

A preferable leader-management technique is called "weak leader renewal". This management technique removes the vigorous upright leader and replaces it with a weaker branch to devigorate the leader, maximise branching, and encourage early fruit production. Weak leader renewal involves pruning the central leader during the dormant season or after bud break, if excessive vigor is a problem, just above the highest usable lateral. The lateral is then pulled up and headed at 12 inches and tied to the tree stake as the new leader. Weak leader renewal can also be used to maintain the leader in ensuing years. Weak leader renewal can also be used with feathered trees that have lower branches and no higher branches. The leader is removed above the highest usable lateral and then that lateral branch is pulled up to replace the leader.

A remedial technique that also may be used for feathered trees with blind wood is to notch above each node in the unbranched region of the leader 2 weeks before bloom with a hacksaw blade going approximately of the way around the tree. Caution should be used so that the cut is only through the bark (phloem) and not the structural wood beneath the bark. Enough of the buds should break in that region, as a result from the notching, to avoid having to head the tree and eliminate the higher branches which would reduce early fruit production.

Another leader-management technique that can be used is called "snaking," which is done during the growing season. Snaking is the process of bending the leader when it is 18 inches long to a 60o angle and taping or tying it to the tree stake. After the leader grows another 18 inches, it is bent back towards the tree at a 60 o angle and taped to the tree stake. Snaking is a technique that can be used to devigorate the leader and increase lateral branching continuously along the leader. In research plots in North Carolina, snaking resulted in greater early fruit production than weak leader renewal or modified heading because the leader was not pruned. Snaking is one technique that can successfully be used to properly train a leader in high density orchards without pruning.

Another technique that has recently been found to be very effective in North Carolina for increasing lateral branching on the unbranched leader, is referred to as "bagging". Bagging is appropriate for the apple grower who will take the time to properly bag trees and remove them in a timely manner. Bagging is done by cutting the unbranched leader to a maximum length of 30 to 36 inches above the highest usable lateral branch 3 to 4 weeks before budbreak. Then, the unbranched central leader is covered with a 2- to 3-mil, small diameter, clear polyethylene bag. Rolls of polyethylene sleeving can be purchased and cut to desired length. The bag is then secured at the top and bottom of the unbranched leader with clothespins or staples. The bag is left on the leader until new lateral growth is 1 to 2 inches long, usually 4 to 5 weeks after bud break. Do not be concerned when the youngest leaves are desiccated several days after removing the bag; trees quickly outgrow this condition. For varieties that are difficult to branch, spray the portion of the leader that was in the bag 3 to 5 days after bag removal with a 250 ppm mixture of Promalin, a plant growth regulator compound of a cytokinin and gibberellins. Promalin will encourage the new shoots to elongate to increase fruit production. For instance, unbranched whips at planting can be bagged to encourage lateral branching instead of heading newly planted trees. Feathered trees with low branches and no high branches can also have the upper portion of the leader bagged without having to prune the tree to encourage lateral branching.

Lateral-Branch Management

When the emerging lateral branches are 3 to 6 inches long, they should be spread outward to develop a wide crotch angle which will form a strong union for future fruit production. Two tools that are effective for spreading branches are toothpicks or spring loaded clothespins. Branches are spread outward as they are 3 to 6 inches long and maintained to approximately an 80 to 85o angle. As the branches grow, the branch can be held down with concrete weights attached to clothespins and clipped to the branch or by tying down with string or twine wrapped loosely around the limb and to the base of the tree stake or clips in the soil. All upright secondary growth on the lateral branches is pulled over to horizontal with weights or pulled over and clipped to other lateral branches or removed when 3 to 4 inches long. Lateral branches that are spread will grow slower due to reduced apical dominance, have more secondary branching and initiate flower buds much earlier due to the reduced vigor. Again, the goal is to avoid pruning and promote early fruit production which is essential for both reducing tree vigor and pulling down lateral branches.

On newly planted trees, all branches with wide crotch angles are left during the first several years to increase initial cropping. If and when shading starts to become a problem in the center of the tree after approximately 3 to 5 years, branches are thinned out to create two distinct whorls of branches approximately 10 to 12 inches apart as mentioned earlier to maximise light penetration.

Lateral branches also need to be controlled in length. During the summer the branches can be cut back to their desired length by cutting into wood that is at least 1 year old to a secondary lateral branch similar in diameter to the branch being cut. If this cut is made during the dormant season, the following spring there will be a vigorous flush of new growth from the are near the dormant cut creating a greater regrowth problem than initially present and will not help in controlling tree size.

With these techniques, a slender, spindle-type high density tree can be developed and maintained in North Carolina. These techniques can be combined to help the apple grower produce slender, spindle-type trees that are both smaller in height for ease of pruning, training and harvesting, and will consistently bear high quality fruit early in the life of the orchard. Although high-density systems have many advantages, summer training and pruning must be conducted approximately every 4 to 6 weeks during the growing season. High-density systems are not like traditional systems which can be "corrected" after a period of neglect. When high-density systems are neglected during the growing season, tree structure, early fruiting, and potential orchard profitability are lost. Because of the greater establishment cost of high-density orchards, a commitment to train the orchard properly is essential. However, the advantages of early high quality fruit production, and an earlier and greater return on your investment are real incentives for high-density orchards.

Fertility

Adequate tree nutrition is essential for quality apple production. Determine the nutrient status of your soil by taking a soil sample prior to planting and each year thereafter at the same time of year. Follow the fertilisation guidelines provided by the soil test. This will prevent over-fertilisation, will be cost-efficient, and will maintain healthy and productive trees. A leaf sample taken in July or August will determine the nutrient status of the tree and can be helpful in conjunction with the soil test. In addition to soil analysis and foliar analysis, regular observation of vegetative growth is a useful indicator of tree fertility. Optimum fertility exists if lateral, outward growth is between 12 and 18 inches per year.

If you are unable to take a soil test, a useful rule is to apply 1 pound of 10-10-10 fertiliser to each tree the first year, 2 pounds the second year, and 3 pounds the third year up to a maximum of 5 to 6 pounds for a mature tree. Always adjust rates of fertiliser application according to annual shoot growth. Apply fertiliser in late winter or early spring before growth begins. Fertiliser should be broadcast on the soil surface around the drip line of the tree. The "drip line" is the circular line at the outer ends of the branches. Avoid getting fertiliser within 6 inches of the trunk as it could burn the tree.

Weed Control

Controlling weeds and gras-ses around young apple trees minimises competition for soil nutrients and moisture, encourages vigorous tree growth, and increases fruit size. It will also help to minimise damage from pine and meadow voles, and other rodents. Keep all vegetation controlled out to the drip line of the tree where the outer branches end. Do not use weed whips; these will girdle the tree without any visible signs of injury. Avoid mechanical weed cultivation, such as tilling, as it damages shallow tree roots. Herbicides are effective, but follow the label directions carefully and keep them off the tree. Mulch will also control weeds and conserve soil moisture, however rodents may burrow under the mulch and gnaw tree trunks and roots.

Pests and Diseases

The insect pests mostly observed are San Jose Scale (*Quadraspidiotus perniciosus*), white scale (*Pseudoulacaspis* sp.), wooly apple aphid (*Eriosoma lanigerum*), blossom thrips (*Thrips rhopalantennalis*) etc. Planting of resistant rootstocks, suitable intercultural operations and spraying with chloropyriphos, fenitrothion, carbaryl etc. have been found to be effective in controlling the pests.

The main diseases reported are collar rot (*Phytophthora cactorum*), apple scab (*Venturia inaequalis*), sclerotius blight (*Sclerotium rolfsii*), crown gall (*Agrobacterium tumefaciens*), cankers, die-back diseases etc. Plants resistant to the diseases should be used for cultivation. The infected plant parts need to be destroyed. Application of copper oxychloride, carbendazim, mancozeb and other fungicides have been found to be effective in controlling the diseases.

In apple, there are three distinct fruit drops, i) early drop resulting from unpollinated or unfertilised blossoms, ii) June drop (due to moisture stress and fruit competition) and iii) Pre-harvest drop. Pre-harvest drop can be controlled by spraying NAA @ 10 ppm. about a week before the expected drop.

Disease and Insect Control

Diseases and insects can cause serious damage to apple trees and fruit. Good sanitation practices are necessary to control pest problems. Cut out all dead or diseased wood, remove dried apples, and clear leaves and fallen debris away from trees. Disinfect pruning tools with a 10% solution of a household disinfectant (Lysol) or bleach, before and after use and between trees. Household disinfectants, such as Lysol, will not corrode tools or ruin clothing. A regular spray program is essential for high fruit quality and healthy trees. Use a multipurpose fungicide and insecticide labeled for apples. These can be obtained from a garden center and will include application instructions. A spray to control fungus problems should be applied when the first sign of green tissue appears. A horticultural oil should also be sprayed on apple trees at the first sign of green growth in the spring to suffocate scale insects and reduce overwintering mite and aphid eggs.

For homeowners with only a few trees, premixed orchard sprays are available from many garden centers. Begin applications after full bloom is over and spray every 10 to 14 days through-out the summer.

Pollination

Apples are self-incompatible; they must cross-pollinate to develop fruit. During the flowering each season, apple growers usually provide pollinators to carry the pollen. Honeybee hives are most commonly used. Orchard mason bees are also used as supplemental pollinators in commercial orchards. Bumble bee queens are sometimes present in orchards, but not usually in enough quantity to be significant pollinators. There are four to seven pollination groups in apples depending on climate:

— Group A - Early flowering, May 1 to 3 in England (Gravenstein, Red Astrachan)

— Group B - May 4 to 7 (Idared, McIntosh)

— Group C - Mid-season flowering, May 8 to 11 (Granny Smith, Cox's Orange Pippin)

— Group D - Mid/Late season flowering, May 12 to 15 (Golden Delicious, Calville blanc d'hiver)

— Group E - Late flowering, May 16 to 18

— Group F - May 19 to 23

— Group H - May 24 to 28

One cultivar can be pollinated by a compatible cultivar from the same group or close (A with A, or A with B, but not A with C or D). Varieties are sometimes classed as to the day of peak bloom in the average 30 day blossom period, with pollinisers selected from varieties within a 6 day overlap period.

Cultivars vary in their yield and the ultimate size of the tree, even when grown on the same rootstock. Some cultivars, if left unpruned, will grow very large, which allows them to bear much more fruit, but makes harvesting very difficult. Mature trees typically bear 40-200 kilograms (88-440 lb) of apples each year, though productivity can be close to zero in poor years. Apples are harvested using three-point ladders that are designed to fit amongst the branches. Dwarf trees will bear about 10-80 kilograms (22-180 lb) of fruit per year.

Commercially, apples can be stored for some months in controlled-atmosphere chambers to delay ethylene-induced onset of ripening. Ripening begins when the fruit is removed. For home storage, most varieties of apple can be stored for approximately two weeks, when kept at the coolest part of the refrigerator (i.e. below 5°C). Some types of apple, including the Granny Smith and Fuji, have an even longer shelf life.

HARVESTING AND FRUIT STORAGE

Apples reach maturity at different times, depending on variety and climate. There is not a specific date at which you can expect to harvest your apples. Instead, you can observe your apples as they grow and inspect the fruit for certain changes which indicate maturity. The "ground" or base skin color of the apples changes from green to yellow as the fruit matures. Flesh color also loses its greenish tint and turns yellow or white. When you are convinced that the apples look mature, take a bite! A mature fruit will be crisp and juicy. A pleasing taste is the final indicator of fruit maturity. Proper storage conditions help prolong the shelf-life of your apples. Store apples at 32 oF and maintain high humidity. The crisper drawers of many refrigerators work well, but keep the fruit away from vegetables since ripening fruit gives off gas that may spoil vegetables. Apples can also be stored in plastic bags in the refrigerator to prevent fruit dehydration.

The orchard start bearing from eighth year and the economic life of an apple tree exceeds 30 years. From eighth to seventeenth year, productivity goes on increasing and thereafter remains constant upto 30 years. The level of productivity varies form elevation to elevation. Production stage extends upto even forty years depending upon agro-climatic condition. Apple being a climacteric fruit, the maturity period does not coincide with ripening. The fruits are usually harvested before they are fully ripe. The average yield of different apple varieties in the state of Uttaranchal is very low (5-6 tonnes/ha.) as compared to that in Himachal Pradesh and Jammu & Kashmir which is around 11-13 tonnes/ha..

Post-Harvest management

Pre-cooling

The fruits are placed in a cool and ventilated place to remove field heat before packing. Fruit surface must be free from moisture before grading, wrapping or packing in cartons.

Grading

Grading of apples is done according to fruit size and fruit appearance or quality. On the basis of fruit size, apples are graded manually in 6 grades. On the basis of fruit colour, shape, quality and appearance, apple fruits can be graded in three or more quality grades. These grades are designated as AAA, AA and A; A,B,C; or extra fancy, fancy class I and fancy class II.

Storage

Apples have a long storage life compared to other fruits and can be stored for a period of 4-8 months after harvesting. The fruits can be kept in cold storage at a temperature of about - 1.10 to 00 C and 85-90% relative humidity.

Packing

Apples are usually packed in wooden boxes having the capacity to accommodate about 10 or 20 kg. fruits. Corrugated fibre board cartons are also used for packing.

Transportation

Road transport by trucks is the most popular mode of transport due to easy approach from orchards to the market.

Marketing

Marketing of the produce is mainly controlled by intermediaries like wholesalers and commission agents. During years of good production, the wholesale prices in producing areas slip down to un-remunerative levels.

APPLE ROOTSTOCKS AND TREE SPACING

The commercial apple industry worldwide is in the midst of a major change in fruit production management systems. With size-controlling rootstocks tree size has been reduced and the number of trees per acre, referred to as tree density, has increased significantly. Some orchards in Europe have exceeded 5,000 trees per acre. However, in North Carolina, tree densities that are being commercially evaluated are around 450 trees per acre with a maximum of approximately 1,100 trees per acre.

All apple trees sold commercially are either budded or grafted trees. The selected variety or top of the tree, referred to as the scion, is grafted or budded on a rootstock of choice. The rootstock determines the relative size of the tree but does not affect the type of fruit or quality of the fruit that the tree bears. With currently available commercial rootstocks, there is a wide range in tree size potential as well as some degree of resistance to certain root borne insects and disease problems.

The terms "dwarf tree" or "dwarfing rootstock" are frequently used to describe trees smaller than full size. However, this term is misleading as a dwarf tree to one person is a medium size tree to the next. It may be best to use the term size-controlling rootstock to avoid misunderstanding.

As a general rule, the smaller the size class of the tree (scion/rootstock) the sooner the tree will bear fruit once planted. It is recommended that trees smaller than M.7a be staked or supported for optimal tree growth. Staking also helps to support the tree and fruit load in the early years as the smaller trees bear fruit earlier. Ten foot stakes are generally recommended with 2 foot driven into the ground and 8 foot above the ground. The tree is loosely tied, approximately every 2 feet, to aid in supporting the leader. Materials commonly used are one inch aluminum conduit or 3 inch wooden posts.

The majority of size-controlling rootstocks originated in England. The "M." prefix, refers to the East Malling Research Station, England, where much of the initial research was conducted in the early 1900's. The "MM." prefix, Malling-Merton, refers to hybrid trees of the Malling series crossed with "Northern Spy" in Merton, England in the 1920's. Some of the "M." and "MM." rootstocks are available with the EMLA suffix, which stands for East Malling/ Long Ashton. The EMLA rootstocks were put through a program to eliminate viral pathogens in the 1960's and are classified as virus-free. Because the trees are virus-free they may be slightly more vigorous than trees on standard rootstocks.

When purchasing apple trees, the tree should be labeled with rootstock and variety in order to correctly space and plant the tree. For example, many apple trees are sold as semi-dwarf trees but without any reference to the rootstock used. Many nurseries define semi-dwarf trees as trees smaller than seedling and larger than the M.7a size class. With a potential difference of 15% in tree size, determining accurate tree spacings for this tree is impossible. Identifying the rootstock when buying your trees will also help to avoid potential failures that are not evident until several years after planting. For example, if the tree was bought as a semi-dwarf on MM.106 and planted in a poorly drained site, the potential for tree loss is very high as MM.106 is not resistant to crown rot.

HIGH-DENSITY APPLE ORCHARD

Commercial apple orchards with trees planted close together on dwarfing or size-controlling rootstocks are referred to as high density plantings. When size-controlling rootstocks are used, tree densities increase from traditional densities of 150 to 250 trees/acre to 500 to 1,000+ trees/acre. Benefits of planting higher-density orchards include earlier production (especially with "fad" varieties); quicker return on investment; training, pruning and harvesting from the ground; potential increased fruit quality; and greater pesticide application efficiency.

When high-density orchards are planted, the training and pruning system must also be modified. For example, characteristics of a typical, full-sized apple tree are: trees planted 20 to 35 ft apart and maintained as a central leader primarily by dormant pruning; height of 15 to 30+ ft with whorls of scaffolds spaced 2 to 3 ft apart on the central leader; and fruit production begins in the 5th to 8th year after planting. Characteristics of size-controlled trees, grown on dwarfing rootstocks are: trees planted 5 to 9 ft apart trained as a miniature central leader with continuous branching along the leader;

maximum height of 8 to 10 ft; and fruit production begins 2 to 3 years after planting. Trees grown on dwarf, size-controlling rootstocks require training and pruning primarily during the summer to control tree size and encourage fruiting early in the life of the tree with minimal dormant pruning. The training techniques discussed in this leaflet are for scion/rootstock combinations, which produce a tree size less than 35% of full size trees.

Many different systems have been proposed for growing size-controlled trees, such as vertical axe, slender spindle, HYTEC, etc. In North Carolina, research has indicated that a slender, spindle-type tree, modified for North Carolina growing conditions, will be most productive. Starting with unbranched, whipped trees, yields of 125 bu/acre in the third year are easily accomplished. If well- branched, feathered trees are planted, fruit production can begin a year earlier and yields of 1,000 bu/acre are attainable in the 5th year.

APPLE CULTIVARS

There are more than 7,500 known cultivars of apples. Different cultivars are available for temperate and subtropical climates. Reputedly the world's biggest collection of apple cultivars is housed at the National Fruit Collection in England. Most of these cultivars are bred for eating fresh (dessert apples), though some are cultivated specifically for cooking (cooking apples) or producing cider. Cider apples are typically too tart and astringent to eat fresh, but they give the beverage a rich flavour that dessert apples cannot.

Commercially popular apple cultivars are soft but crisp. Other desired qualities in modern commercial apple breeding are a colourful skin, absence of russeting, ease of shipping, lengthy storage ability, high yields, disease resistance, typical "Red Delicious" apple shape, long stem, and popular flavour. Modern apples are generally sweeter than older cultivars, as popular tastes in apples have varied over time. Most North Americans and Europeans favour sweet, subacid apples, but tart apples have a strong minority following. Extremely sweet apples with barely any acid flavour are popular in Asia and especially India.

Old cultivars are often oddly shaped, russeted, and have a variety of textures and colours. Some find them to have a better flavour than modern cultivators, but may have other problems which make them commercially unviable, such as low yield, liability to disease, or poor tolerance for storage or transport. A few old cultivars are still produced on a large scale, but many have been kept alive by home gardeners and farmers that sell directly to local markets. Many unusual and locally important cultivars with their own unique taste and appearance exist; apple conservation campaigns have sprung up around the world to preserve such local cultivars from extinction. In the United Kingdom old cultivars such as Cox's Orange Pippin and Egremont Russet are still commercially important even though by modern standards they are low yielding and disease prone.

REFERENCES

Billy, Simon. *Horticultural Farm Manager*. The Kerr Center for Sustainable Agriculture, Poteau, OK. Personal communication. May 1997.

Bordelon, B. 2002. *Grape Cultivars for the Upper Midwest*. Purdue University, Department of Horticulture and Landscape Architecture.

Chan, h.t. Jr. 1984. (Ed.). *Handbook of Tropical Foods*. New York/Basel: Marcel Dekker.

Chandler, William Henry. 1958. *Evergreen Orchards*. Philadelphia: Lea and Febiger.

3

Apricot

The Apricot is a species of Prunus, classified with the plum in the subgenus Prunus. The native range is somewhat uncertain due to its extensive prehistoric cultivation, but most likely is northern and western China and Central Asia and possibly also Korea and Japan.

It is a small tree, 8-12 m tall, with a trunk up to 40 cm diameter and a dense, spreading canopy. The leaves are ovate, 5-9 cm long and 4-8 cm wide, with a rounded base, a pointed tip and a finely serrated margin. The flowers are 2-4.5 cm diameter, with five white to pinkish petals; they are produced singly or in pairs in early spring before the leaves. The fruit is a drupe similar to a small peach, 1.5-2.5 cm diameter, from yellow to orange, often tinged red on the side most exposed to the sun; its surface is usually pubescent. The single seed is enclosed in a hard stony shell, often called a "stone", with a grainy, smooth texture except for three ridges running down one side.

The Apricot was first cultivated in India in about 3000 BC. In Armenia it was known from ancient times; having been brought along the Silk Road; it has been cultivated there so long it is often thought to be native there. Its introduction to Greece is attributed to Alexander the Great, and the Roman General Lucullus also exported some trees, cherry, white heart cherry and apricot from Armenia to Europe. Subsequent sources were often much confused over the origin of the species. Loudon believed it had a wide native range including Armenia, Caucasus, the Himalaya, China and Japan. Nearly all sources presume that because it is named armeniaca, the tree must be native to or have originated in Armenia as the Romans knew it. For example, De Poerderlé asserts: "Cet arbre tire son nom de l'Arménie, province d'Asie, d'où il est originaire et d'où il fut porté en Europe ...." . There is no scientific evidence to support such a view. Today the cultivars have spread to all parts of the globe with climates that support it.

Apricots have been cultivated in Persia since antiquity, and dried ones were an important commodity on Persian trade routes. Apricots remain an important fruit in modern-day Iran where they are known under the common name of *Zard-ālū.*

More recently, English settlers brought the apricot to the English colonies in the New World. Most of modern American production of apricots comes from the seedlings carried to the west coast by Spanish missionaries. Almost all U.S. production is in California, with some in Washington and Utah.

Figure 1. Apricot fruits

Many apricots are also cultivated in Australia, particularly South Australia where they are commonly grown in the region known as the Riverland and in a small town called Mypolonga in the Lower Murray region of the state. In states other than South Australia apricots are still grown, particularly in Tasmania and western Victoria and southwest New South Wales, but they are less common than in South Australia.

CULTIVATION

Although often thought of as a "subtropical" fruit, this is actually false - the Apricot is native to a continental climate region with cold winters, although can grow in Mediterranean climates very well. The tree is slightly more cold-hardy than the peach, tolerating winter temperatures as cold as -30 °C or lower if healthy. The limiting factor in apricot culture is spring frosts: They tend to flower very early, around the time of the vernal equinox even in northern locations like the Great Lakes region, meaning spring frost often kills the flowers. Furthermore, the trees are sensitive to temperature changes during the winter season. In their native China, winters can be very cold, but temperatures tend to be more stable than in Europe and especially North America, where large temperature swings can occur in winter. The trees do need some winter cold to bear and grow properly and do well in Mediterranean climate locations since spring frosts are less severe but there is some cool winter weather to allow a proper dormancy. The dry climate of these areas is best for good fruit production. Hybridisation with the closely related *Prunus sibirica* offers options for breeding more cold-tolerant plants.

Apricot cultivars are most often grafted on plum or peach rootstocks. A cutting of an existing apricot plant provides the fruit characteristics such as flavour, size, etc., but the rootstock provides the growth characteristics of the plant. Apricots and plums can hybridise with each other and produce fruit that are variously called pluots, plumcots, or apriums.

There is an old adage that an apricot tree will not grow far from the mother tree. The implication is that apricots are particular about the soil conditions in which they are grown. They prefer a well-drained soil with a pH of 6.0 to 7.0. If fertiliser is needed, as indicated by yellow-green leaves, then 1/4 pound of 10-10-10 fertiliser should be applied in the second year. Granular fertiliser should be scattered beneath the branches of the tree. An additional 1/4 pound should be applied for every year of age of the tree in early spring, before growth starts. Apricots are self-compatible and do not require pollinizer trees, with the exception of the 'Moongold' and 'Sungold' cultivars, which can pollinate each other. Apricots are susceptible to numerous bacterial diseases including bacterial canker and blast, bacterial spot and crown gall. They are susceptible to an even longer list of fungal diseases including brown rot, Alternaria spot and fruit rot, and powdery mildew. Other problems for apricots are nematodes and viral diseases, including graft-transmissible problems.

PRODUCTION TRENDS

Turkey (Malatya region) is the leading apricot producer, followed by Iran. In Armenia apricots are grown in Ararat Valley. Details of global apricots production is summarised in table below.

Top Ten Apricot Producers - 2005

(1,000 tonnes)

Country	Production
Turkey	390
Iran	285
Italy	232
Pakistan	220
Greece	196
France	181
Algeria	145
Spain	136
Japan	123
Morocco	103
Syria	101
World Total	1916

Seeds or kernels of the apricot grown in central Asia and around the Mediterranean are so sweet that they may be substituted for almonds. The Italian liqueur Amaretto and amaretti biscotti are flavoured with extract of apricot kernels rather than almonds. Oil pressed from these cultivars has been used as cooking oil.

MEDICINAL AND NON-FOOD USES

Cyanogenic glycosides are found in high concentration in apricot seeds. Laetrile, a purported alternative treatment for cancer, is extracted from apricot seeds. As early as the year 502, apricot seeds were used to treat tumors, and in the 17th century apricot oil was used in England against tumors and ulcers. However, in 1980 the National Cancer Institute in the USA described laetrile to be an ineffective cancer treatment.

In Europe, apricots were long considered an aphrodisiac, and were used in this context in William Shakespeare's *A Midsummer Night's Dream,* and as an inducer of childbirth, as depicted in John Webster's *The Duchess of Malfi.*

Due to their high fiber to volume ratio, dried apricots are sometimes used to relieve constipation or induce diarrhea. Effects can be felt after eating as few as three.

Research shows that of any food, apricots possess the highest levels and widest variety of carotenoids. Carotenoids are antioxidants that help prevent heart disease, reduce "bad cholesterol" levels, and protect against cancer. In traditional Chinese medicine, apricots are considered helpful in regenerating body fluids, detoxifying, and quenching thirst.

Some claim that the kernels also have healthy properties, including toning the respiratory system and alleviating a cough. However, the tip of the apricot holds a concentrated amount of the chemical laetrile, which can be upsetting to the system. The tips of the seeds should be removed and consumption should be limited to no more than five a day.

APRICOT VARIETIES

The best place to start with apricots is sampling them fresh out of hand. Select apricots that are golden orange and plump, not too soft and not too hard. Apricots that are soft and ripe will have the best flavor. Give them the taste test immediately.

Fresh apricots come to market from mid-spring to mid-summer, May through July and even later in the Northern Hemisphere. The apricot harvest can be divided into early, mid-season and late. That means you can enjoy some or all of these over the course of the apricot harvest. Here are a few apricot varieties to look for:

Apium: medium to large fruit with clear yellow skin and some plum taste. Late harvest.

August Glo: medium-size apricot with a great sweet-tart flavor. Late harvest.

Autumn Royal: similar to Blenheim; medium fruit with yellow skin with a slight orange blush; flesh is yellow, slightly acid. Use fresh, canned, or dried. Late harvest into fall.

Blenheim: medium to large apricot with thick yellow-orange flesh, very juicy fruit with sweet, sprightly, and aromatic flavor. This is the classic California apricot. Eat out of hand or use for canning. Early to midseason harvest.

Chinese (Mormon): small, orange-skinned fruit with red blush. Smooth, firm flesh is sweet and juicy. Midseason to late harvest.

Earligold: medium-sized fruit is golden yellow with a rich and juicy flesh. Use for canning and for eating fresh. Early harvest.

Floragold: small to medium size apricot with yellow skin and flesh. Mid-season harvest.

Garden Annie: medium to large fruit with a bright yellow skin; the clingstone flesh is juicy and firm. Early harvest.

Gold Kist: medium to large apricot with a red-blushed yellow skin. Excellent sweet-tart flavor. Use for canning, freezing, drying and for eating fresh. Early harvest.

Goldcot: medium too large fruit nearly round with bright gold skin; thick flesh is orange and firm and sprightly-sweet flavored. Use for processing, canning, or eating fresh. Midseason to late harvest.

Golden Amber: large and symmetrical apricot with yellow skin and flesh, firm, slightly acid. Use fresh, in canning, or for drying. Late harvest.

Harcot: medium to large fruit, firm, sweet, and juicy. Early harvest.

Harglow: medium-sized orange apricot, sometimes blushed red; firm, sweet and flavorful. Early harvest.

Hargrand: very large, firm apricot. Midseason harvest.

Harogem: good flavor. Late harvest.

Katy: large apricot with red-blushed skin and deep yellow flesh; freestone flesh is firm mild and sweet. Early harvest.

King: very large apricot. Midseason harvest.

Manchurian apricot: small, mild-flavored, orange fruit. Use for drying. Early harvest.

Montrose: large apricot with yellow skin and red blush; sweet, juicy great flavor; edible kernel. Late harvest.

Moongold: plum-size apricot with golden skin and flesh; sweet, sprightly flavor. Use fresh, canned or for jam. Early to midseason harvest.

Moorpark: this large apricot was developed in the eighteenth century. Considered one of the best. Juicy and aromatic; sweet rich, plum-like taste. Brownish-red skin with specks and dots; yellow to orange flesh. Midseason harvest.

Newcastle: small to medium round well-shaped apricot with lemon-yellow skin; flesh can be soft and coarse. Midseason harvest.

Perfection (Goldbeck): fruit is large, oval to oblong, light yellow-orange skin with a pebbly appearance; yellow to yellow-orange flesh; mediocre flavor. Early harvest.

Plum Parfait: medium fruit; skin is red blushed over dark yellow; flesh is dark yellow marbled red at the pit. This is a hybrid of a Japanese plum and apricot. Early harvest.

Puget Gold: medium size apricot with good flavor low in acid. Use canned or dried. Mid- to late-season harvest.

Riland: nearly round, medium fruit is covered with fine velvety hairs; light yellow skin has deep red blush over half of the fruit; the flesh if firm and meaty. Ripens from the pit out. Midseason harvest.

Rival: large, oval orange apricot blushed red. Early harvest.

Royal Rosa: medium fruit with bright yellow skin; firm flesh has pleasant aroma; the flavor is a good balance between sugar and acid. Best for eating fresh. Midseason to late harvest.

Royalty: extra-large apricot. Early harvest.

Sungold: plum-size, bright orange apricot with a sweet, mild flavor. Use fresh, canned, or as jam. Early to midseason harvest.

Sun-Glow: very colorful fruit. Midseason harvest.

Tilton: large to very large apricot with orange skin yellow-orange flesh; fair flavor. Use fresh. Midseason harvest.

Tomcot: large, sweet fruit with sweet orange flesh. Early harvest.

Wenatchee Moorpark: Large to oval apricot with orange-yellow flesh and skin; fair texture. Excellent flavor. Midseaon harvest.

GROWING APRICOTS

Young apricot trees are likely to make excessive growth, especially trees from 2 to 5 years of age. Trees making excessive growth do not mature early in the fall and are, therefore, more subject to injury from low temperatures in November and early winter. Mature apricot trees are hardier than young trees.

Trees should not make more than 2 feet of annual growth.

It is extremely important to train young trees correctly to prevent weak crotches that are very susceptible to winter injury.

The southwest side of the trunk can become very warm on sunny days in late fall and winter. Night temperature frequently drops to well below freezing. This alternative freezing and thawing often injures the bark and wood. This permits wood-destroying fungi to gain entrance, and a large canker is likely to develop.

Paint the trunk with a latex-based, white paint. This paint is non-caustic because it is fast-drying and does not contain turpentine or oil. It gives good

coverage throughout the winter, and the white surface reflects the sun's rays, preventing deep bark penetration. Brush the paint onto the trunk from the ground line up to the first scaffold branch.

Caution: Do not use ordinary house paintts containing oil, turpentine or lead. These may seriously injure or kill the trees.

Fill any holes that develop in the soil at the base of the trunk. Make certain that the soil level adjacent to the trunk is slightly higher so that water will drain away from the tree. Water collecting in depressions near the trunk will form ice in late fall or winter. This ice may girdle the tree, causing death or serious injury.

ROOTSTOCKS

Do not use apricot trees grafted on peach seedlings. Manchurian apricot seedlings and seedlings of the Michigan apricot selection, Goldcot, are compatible with apricot varieties. The Manchurian seedlings are es[illegible]lly winter hardy and have given strong, healthy trees with minimum tree loss.

Varieties

The standard varieties produced in the West have failed in Michigan. Goldcot is the primary variety grown in Michigan. The cultivars listed are reported to be hardy in the north.

— *Goldcot* - The most dependable variety for planting in Michigan.

— *Early Golden Hardy Iowa* - Good for fresh fruit or cooking.
— *Moongold* - Must be planted with Sungold for pollination. Fruit suitable for most uses.
— *Sungold* - Must be planted with Moongold. Good fresh or in preserves.

Use of Pollinators

Goldcot is apparently completely self-fruitful and can be planted alone. The other varieties, while ordinarily self- fruitful, will set heavier crops when unfavorable weather prevails during blossoming periods if they are inter-planted with other fertile varieties.

Planning Tips

New trees should be planted in the early spring. They can be planted in the fall in mild areas if the trees are dormant. When selecting trees, look for trees that are least 1 year in age. Trees are self-fertile (they do not need other trees nearby to cross pollinate with in order to produce fruit), but tend to produce more successfully if planted near other varieties. Trees begin producing fruit when they are 3 to 4 years old- 3 to 4 bushels from standard varieties and 1 to 2 bushels from dwarf varieties.

Site and Soil

Apricot trees need a site with full sun, and can tolerate a wide range of soils as long as the soil is well-drained. Trees grown in northern regions benefit from being planted on the north side of buildings so trees stay cool as late as possible into the spring.

Apricot trees are sensitive to climatic conditions and require the best possible growing sites to remain healthy and regularly productive. Some favorable Michigan sites for apricots exist in the northwestern area (Leelanau county) of Michigan. Apricots seem to grow and produce well in locations where sweet cherries are productive.

Apricots blossom about a week earlier than peaches. Therefore, plant apricots only on sites which are practically frost-free. Elevation of the site above the surrounding country will give added protection against loss from frost and winter injury. Nearness to a large body of water is associated with slow warming temperatures in the spring, which help hold back spring bud development.

The soil must be well drained and preferably of a sandy type. Poor subsoils of any kind will result in the death or poor growth of many trees. Avoid heavy soils for apricots as such soils are likely to be poorly drained.

Planting

Standard-sized apricot trees should be planted at least 25 feet from other trees. Dwarf varieties can be spaced between 8 and 12 feet apart. Keep an area about 4 feet in diameter free from weeds and grass during the early stages of growth. Mulch newly planted trees with several inches of organic material, leaving 3 to 4 inches around the base of the trunk free of mulch.

Fall-planted trees are more subject to cold injury during the first winter after planting. Plant the bud union 4 inches below ground level for good anchorage. The soil should be well settled and firm around the roots.

Apricot trees grow to become relatively large. An area with a minimum diameter of 25 ft. (7.6 meters) should be allowed. Select a well grown one or two year old tree from the nursery. Two year old trees should have at least four of five well-spaced branches, with a good root system. The usual practice is to plant early in the spring, but planting can be completed in the fall when weather conditions are good and the soil is moist.

Prepare a hole slightly larger than the root spread. Trim off any broken roots before planting. If the tree is in a plastic pot, remove the pot. If it comes in a fiber pot, you can slit the sides and plant with the pot or remove the pot. Sprinkle a handful of bone meal (phosphorus) in the bottom of the hole to help the root system get established. Place the tree in the hole. Mix in some peat moss or compost with the planting soil. Replace the soil in the hole, treading the soil firmly around the roots to ensure that the tree is securely anchored in the ground. Give the tree a good watering. An area of about 4 ft. in diameter (1.2 meters) should be kept free of weeds or lawn grass during the early stages of growth. Organic or plastic mulches can also be used to suppress weed growth around the tree.

Care and Maintenance

Trees will benefit from an application of nitrogen fertiliser in the spring. Young trees should be trained to have an open center (leader trunk removed), with many well-spaced branches that are capable of bearing their heavy fruit without breaking. Shoots and suckers should be pruned annually to encourage the growth of fruiting spurs. If the tree sets a crop of fruit and no thinning is done, the overall size of the fruit will be reduced when it comes time to harvest. When fruits reach 1 inch in diameter they should be thinned to 3 to 4 fruits per cluster to increase the overall size of the remaining apricots. Watch for common fruit pests and diseases in the spring and fall.

Training

Train the young tree to the modified central-leader system. Select two main scaffolds on opposite sides of the trunk, well separated so they will not touch each other when fully grown. The first scaffold should be about 30 inches above the ground if trees are to be mechanically harvested. The second scaffold should be 8 to 10 inches above the first. Select scaffolds having wide angles at the point of attachment with the trunk.

Sharp-angled branches split badly. The leader must be left longer than the scaffolds so that it will not be shaded out.

It is very important the second spring to remove all excess scaffolds from the trunk, retaining just two or three. Head the scaffolds back somewhat if they are likely to grow higher than the leader. By the third and fourth years, the trees are beginning to produce some fruit. The main framework of the tree has been established, and pruning is mostly to remove additional branches that may come from the trunk, and to head back new terminal growth if it exceeds 2 feet in length. A small amount of thinning out may be necessary where branches are too thick or are rubbing. Do not leave too many branches originating close together on the trunk.

At planting time cut a one year old tree back to a height of 33 to 36 inches (82.5-90 cm). If a two year old tree is planted, reduce the branches to four well spaced shoots and shorten each one by one third. Apricots are usually grown as open center trees with the central leader removed. Aim to develop a framework of well-spaced branches that are capable or bearing heavy crops without breaking. In subsequent years build up the framework branches and cut out the entire shoots that are crowded or crossing into the center of the tree. Narrow angled crotches should be avoided as these are sources of weakness. Because apricot trees can grow to be large, in later years it may be necessary to cut or head back limbs in order to encourage more growth in the lower parts of the tree. Always cut back to a lateral or side growing branch.

Pruning

The apricot produces most of its fruit on rather short- lived spurs. Prune mature trees to remove branches loaded with old spurs and keep the trees producing good replacement wood. This requires a combination of thinning out branches having many weak spurs and heading back long branches by one-third or to a strong lateral branch. Mature trees must make from 16 to 24 inches of new terminal growth a year to maintain satisfactory annual production. If less growth is made, the trees will fall into a biennial bearing habit and produce a large crop of small apricots every other year. Prune or head back the long branches in April. Additional pruning in June and August will reduce excess vigor. Do not prune during late fall or mid-winter because of drying out and the possibility of winter injury.

Grow young apricot trees slowly. They have a tendency to make excessive growth, rendering them more susceptible to winter injury. New growth should not exceed 2 feet in length annually.

Nitrogen is the most important nutrient in fruit production and is usually the only fertiliser element that should be applied regularly in Michigan. However, young apricot trees have great inherent vigor even on relatively poor soils; therefore, nitrogen fertilisers should not be used in young plantings on reasonably fertile land and sparingly even on light, sandy soils. As the trees become older, start producing and the danger of winter injury becomes less, nitrogen applications will need to be increased sufficiently to insure an annual terminal growth of 16 to 24 inches to maintain good production. Apply nitrogen in late fall, after mid-November, or before growth starts in spring. The kind of nitrogen to use should be chosen on the basis of cost of actual nitrogen and ease of application; however, nitrogen in the form of urea should be avoided.

Potassium is the only nutrient element, other than nitrogen, that is likely to be needed in Michigan apricot plantings. Potassium deficiency reduces tree

growth, yields and fruit quality. The best way to determine need for potassium (potash) is through leaf analysis. When needed, potash may be applied in the fall or spring. The most common potassium fertiliser is muriate of potash (60 percent K_2O).

Phosphorus is utilised in only small amounts by fruit trees compared to either nitrogen or potassium. There is no present indication that apricot trees will benefit from application of phosphorus. However, in some cases, fertilisers containing phosphorus may benefit growth of cover crop.

Fruit Thinning

If the tree sets a heavy crop and no thinning is done, the fruit will be small at harvest time. Thus, removal of part of the crop is necessary. To do this, space the fruits about 1 - 2 inches (3.8 - 5 cm) apart. Early thinning results in more uniform ripening. Fruit on well-thinned trees will ripen several days earlier than on poorly or unthinned trees. There will still be mixed maturities, even on well thinned trees, so more than one pick may be necessary. Heat greatly accelerates maturity.

If fruit set is heavy, thin the individual fruits so that they will be 1 1/2 to 2 inches apart. Otherwise, fruits will be small, and the trees will become biennial in bearing habit. Thinning can be done by hand, or — faster and easier — by careful pole thinning. Bamboo poles of various lengths are used. Cover the top 12 inches with a piece of garden house to reduce injury to the branches. The excess fruits may be removed by rubbing and tapping. Avoid hard blows, as all of the fruits will be knocked off for a considerable distance from the point of contact. Early thinning is recommended. Start thinning as soon as danger of frost has passed.

Soil and Fertiliser

Soils in the Southern Interior are chronically low in organic matter and nitrogen. Minor elements such as magnesium, boron, and zinc may be low as well. If good weed control is practiced, no fertiliser should be required for the first two or three years. When the tree starts to crop, apply one ounce (28 grams) of a complete fertiliser such as 12-16-12 (which contains minor elements) per square yard (0.8 sq. meters) in the fall. In mature trees the aim is to get 15 inches (38 cm) of new growth every year. Nutrients can also be applied as foliar sprays in early summer. Organic growers should use approved sources of organic nutrients.

Harvesting

Apricots must mature on the tree but they can ripen either on or off the tree. If left until they are good to eat, they will bruise very easily with picking and transport. Background color is used as a guide for harvest maturity.

Apricots are harvested before fruit has fully ripened on the tree. The fruit is ready when it has attained a uniform color and the skin gives slightly when pressed. Harvest season peaks in July or August depending on zone. For best taste, apricots should be allowed to ripen fully before use.

REFERENCES

Jackson, David I., and Norman E. Looney. 1999. *Temperate and Subtropical Fruit Production.* 2d ed. Wallingford, Oxon., U.K.: CABI.

Popenoe, Wilson. *Manual of Tropical and Subtropical Fruits.* 1920. New York: Hafner Press.

Ryugo, Kay. 1988. *Fruit Culture: Its Science and Art.* New York: Wiley.

Roach, F. A. 1985. *Cultivated Fruits of Britain.* Oxford: Blackwell.

4

Cherry

The word cherry refers to a fleshy fruit that contains a single stony seed. The cherry belongs to the family Rosaceae, genus *Prunus,* along with almonds, peaches, plums, apricots and bird cherries. The subgenus, Cerasus, is distinguished by having the flowers in small corymbs of several together, and by having a smooth fruit with only a weak groove or none along one side. The subgenus is native to the temperate regions of the Northern Hemisphere, with two species in America, three in Europe, and the remainder in Asia.

The cherry is generally understood to have been brought to Rome from northeastern Anatolia, historically known as the Pontus region, in 72 BC. The city of Giresun in present-day Turkey was known to the ancient Greeks as Choerades or Pharnacia, and later as Kerasous or Cerasus, < Kerason < Kerasounta < Kerasus "horn" (for peninsula) in Greek + *ounta* (Greek toponomical suffix). The name later mutated into *Kerasunt* (sometimes written *Kérasounde or Kerassunde*).

The English word cherry, French cerise, Spanish cereza, and Southern Italian dialect cerasa all come from the Classical Greek through the Latin cerasum, thus the ancient roman place name Cerasus, from which the cherry was first exported to Europe. By the Middle Ages, however, cherries had disappeared in England. They were reestablished at Tyneham, near Sittingbourne in Kent by order of Henry VIII, who had tasted them in Flanders.

Cherries contain anthocyanins, the red pigment in berries. Cherry anthocyanins have been shown to reduce pain and inflammation in rats. Anthocyanins are also potent antioxidants under active research for a variety of potential health benefits. According to a study funded by the Cherry Marketing Institute presented at the Experimental Biology 2008 meeting in San Diego, rats that received whole tart cherry powder mixed into a high-fat diet did not gain as much weight or build up as much body fat, and their blood showed much lower levels of inflammation indicators that have been linked to heart disease and diabetes. In addition, they had significantly lower blood levels of cholesterol and triglycerides than the other rats.

Figure 1. Branch of a cherry plant

The Wild Cherry (*P. avium*) has given rise to the Sweet Cherry, to which most cherry cultivars belong, and the Sour Cherry (*P. cerasus*), which is used mainly for cooking. Both species originate in Europe and western Asia; they do not cross-pollinate. The other species, although having edible fruit, are not grown extensively for consumption, except in northern regions where the two main species will not grow. Irrigation, spraying, labor and their propensity to damage from rain and hail make cherries relatively expensive. Nonetheless, there is high demand for the fruit.

Cherries have a very short growing season and can grow anywhere, including the great cold of the tundra. In Australia they are usually at their peak around Christmas time, in southern Europe in June, in North America in June, and in the UK in mid July, always in the summer season. In many parts of North America they are among the first tree fruits to ripen.

Besides the fruit, cherries also have attractive flowers, and they are commonly planted for their flower display in spring; several of the Asian cherries are particularly noted for their flower displays. The Japanese sakura in particular are a national symbol celebrated in the yearly Hanami festival. Many flowering cherry cultivars have the stamens and pistils replaced by additional petals, so are sterile and do not bear fruit. They are grown purely for their flowers and decorative value. The most common of these sterile cherries is the cultivar "Kanzan".

CLASSIFICATION

This refreshing and wholesome fruit is by no means a modern development. In the fourteenth century we are told by Marco Polo and others, that good cherries were grown in the kingdom of Timur the Great in Asia. Koch in his "Dendrologie" speaks of wild and cultivated cherries - both sweet and sour - over the parts of Asia Minor he visited, and Dr. E. Regel, who lived in Turkestan nine years, says: "Stately trees of the sweet cherry stand near Kâratag in the Hisser district. The original district for cultivation of the sour cherries in central Asia embraces all of west Turkestan from Tashkend and Kokham to the upper Amudaria and Afghanistan. Well-flavored, clear, red kinds are found in Baldshuan. In Shugnan the cherry juice is used as a cooling drink, but cherry brandy is unknown." At the great commercial fair at Nishni Novgorod in 1883 the writer and Mr. Charles Gibb were told by Asiatic merchants and traders that tall-growing sweet cherries and nearly sweet cherries of the Vladimir type were grown up to the 56th parallel of north latitude and eastward to the Ural Mountains, where water was obtainable. Botanically it is true that all cultivated cherries have originated, as stated by De Candolle, from two species which are yet found wild:

1. Seedling Variations

Prunus avium is tall, with no suckers from the roots, leaves downy on the under side, and the fruit sweet.

2. Seed Variation of Cultivated Plants

Prunus cerasus, shorter in growth, with suckers from the roots, leaves glabrous, and fruit more or less sour or bitter.But in the centuries of evolution in Asia and east Europe, as well as west Europe, varieties originating from the two original species have been crossed and recrossed until any attempt to classify them botanically must prove unsatisfactory. On this subject Dr. Hooker says: "As with most plants which have been long cultivated it is a matter of difficulty, if not an impossibility, to identify the parent stock of the numerous cultivated varieties of the cherry." Robert Hogg, the celebrated English pomolo-gist, gave many years of study to a horticultural classification of the European varieties, and in his "Fruit Manual," he gives eight classes which are divided into eight races, as follows:

Class I. Geans

Branches rigid and spreading, forming round-headed trees. Leaves long, waved on the margin, thin and flaccid, and feebly supported on the foot-stalks. Flowers large and opening loosely, with thin, flimsy obovate or roundish ovate petals. Fruit heart-shaped, or nearly so. Juice sweet.

Race 1. Black Oeans.—Fruit obtuse, heart-shaped; flesh tender and melting; dark; juice uncolored. Black Eagle and early purple gean are examples of this class.

Race 2. Red Geans—Flesh pale; juice uncolored. Such as Downer's Late and Early Amber.

Race 3. Black Hearts.—Fruit heart-shaped; flesh half firm; dark; juice colored. Examples are Black Heart and Black Tartarian.

Race 4. Red Hearts or Bigarreaus.—Flesh pale; juice uncolored. Examples, Governor Wood and Elton.

Class II. Griottes

Branches spreading at upright, or more or less long, slender, and drooping; leaves flat, dark green, glabrous beneath and borne stiffly on the leaf-stalks, large and broad in Class I and narrow in Class II; flowers in pedunculate umbels, cup-shaped, with firm and crumpled orbicular petals; fruit round or oblate and sometimes in the Morello heart-shaped ; juice subacid or acid.

Race 1. Black Dukes. - Branches upright, occasionally spreading; leaves large and broad; flesh dark; juice colored. Examples, Empress Eugenia and May Duke.

Race 2. Red Dukes. - Flesh pale; juice uncolored. Examples, Belle de Choise, Carnation, and Late Duke.

Race 3. Black Morellos. - Branches long, slender, and drooping; leaves small and narrow; flesh dark; juice colored. Examples, Double Natte, Cerise de Ostheim, and English Morello.

Race 4. Red Morello or Kentish. - Flesh pale; juice uncolored. Examples, Early Richmond, Early Red, Late Kentish, and Montmorency.

With this system in our vest-pockets, the writer and Mr. Charles Gibb in 1882 were able to classify all cherries in their season of fruitage in going eastward from France through Germany, Austria, Poland, and Russia to the Volga. The names of the sour and sometimes the duke varieties were often misleading, but the Bunt Amarelle was plainly a Red Duke, as was also the Lutovka and the Black Amarelles, and the Weichsels with colored juice were plainly Morellos and those with uncolored juice and red color were as plainly Red Morellos or Kentish. It is true that such east European varieties as Brüsseler Braune and Vladimer with colored skin and juice differ in tree habit and texture of fruit from the English Morello, yet they do not differ as much as the apples or oranges that are classed together, and the same is true of all the Red Morellos of all parts of Europe with red skin and uncolored flesh and juice. After separating into two groups the main after differences in the Morellos is in hardiness of tree as grown in various parts of the Union and difference in quality for varied uses.

COMMERCIAL ORCHARDING

Annual world production of domesticated cherries is about two million tonnes. Around 40% of world production originates in Europe and around 13% in the United States. The US is the world's second largest single country producer, after Turkey.

Major commercial cherry orchards in Europe extend from the Iberian peninsula east to Asia Minor, and to a smaller extent may also be grown in the Baltic States and southern Scandinavia.

In the United States, most sweet cherries are grown in Washington, California, Oregon, and Northern Michigan. Important sweet cherry cultivars include "Bing", "Brooks", "Tulare", "King" and "Rainier". Both Oregon and Michigan provide light-colored "Royal Ann" cherries for the maraschino cherry process. Most sour cherries are grown in Michigan, followed by Utah, New York, and Washington. Additionally, native and non-native cherries grow well in Canada. Sour cherries include Nanking and Evans Cherry. Traverse City, Michigan claims to be the "Cherry Capital of the World", hosting a National Cherry Festival and making the world's largest cherry pie. The specific region of Northern Michigan that is known the world over for tart cherry production is referred to as the "Traverse Bay" region. Farms in this region grown many varieties of cherries, sold through companies in the region.

In Australia, the New South Wales town of Young is famous as the "Cherry Capital of Australia" and hosts the internationally famous National Cherry Festival. Popular varieties include the "Montmorency", "Morello", "North Star", "Early Richmond", "Titans", and "Lamberts".

Cherry cultivation is gaining popularity in Himachal Pradesh as a profitable alternative to other cash crops. Prunus avium, our common cherry, is a delicious fruit, rich in protein, sugar and minerals and has more calorific value than apple. It is grown in areas between 2,000 and 2,700 meters above sea-level, requiring 1,000-1,500 hours of chilling period during winters and the climate of Himachal Pradesh is most suited for its growth, pointed out JS Chandel of Dr YS Parmar University of Horticulture and Forestry Nauni, Solan.

With over 2,000 hectares of land under cherry cultivation and production of 438 tonne in 2008, Himachal Pradesh that has emerged as the second largest cherry producing state in the country after Jammu and Kashmir. Of the total, about 420 tonne was grown in Shimla region alone in the cherry belts of Rampur, Narkanda, Rohru ,Kandayli, Kotgarh and Kotkhai.

One of the reasons for the popularity of cherries is that it provides supplementary income to growers as September is the month for apples while cherry blossoms in May-June. As cherry crop is ready much before its time for apple to blossom, cherry farming provided employment to 16 lakh people in the state last year, in marketing, packaging and direct sale to tourists.

Women, particularly in rural Himachal, have taken to cherry farming and marketing to supplement their family earning. The price of cherry at the Chandigarh fruit and vegetables market in Sector 26 is between Rs 250 and Rs 300 for a box of one kilogram.

Himachal Pradesh produced 698 tonne of cherry in 2007, but its production slumped to 455 tonne in 2008 due to dry weather. This year, cherry production is expected to cross the 700-tonne mark in view of favourable weather conditions.

Cherry trees grow well in most zones. Although if you want to get more technical, tart cherries thrive best in zones 4-9, and sweet cherries grow best in zones 5-9. Bush cherries are hardy enough for zone 3. Thanks to modern plant breeding, there are now self-pollinating cultivars and dwarf root-stocks of both tart and sweet cherry trees. Whether your preference is tart or sweet, select a healthy, disease-resistant tree cultivated for your zone. Expect standard-sized trees to start bearing fruit in their fourth year (30 to 50 quarts), and dwarf-sized trees to produce fruit in their third year (10 to 15 quarts).

Site Preparation

Plant trees or bushes in a sunny site with good air circulation and average to rich, well-drained soil. They prefer a soil pH of between 6.0 and 6.8 (slightly acidic). Avoid low areas where frost and standing water can be a problem (especially for sweet types) or sites where cherries, peaches or plums have grown previously. Wild choke-cherries should be located well away from your intended site.

Planting

Cherry trees should be planted in the early spring (zones 4-6) or fall. All purchased cherry trees come grafted on a rootstock, and the type of rootstock it's grafted on will determine its performance, size and how deep it's planted. Consult with the nursery to determine exact planting specifications for the type you buy. Generally speaking, when spacing trees for planting, allow for a distance of 20 to 30 feet in every direction around a standard-sized sweet cherry tree; dwarf trees 8 to 12 feet. Tart cherry trees should be planted about 15 to 20 feet apart. Mulch trees around the base, leaving 4-5 inches next to the trunk bare.

Care and Maintenance

Semi-dwarf and standard-size sweet cherry trees should be trained to have a central leader shape. Tart cherry trees seem to respond better to a modified leader structure. Trees should be pruned annually in late winter while they are in their dormant state. Apply fertilizer in the spring until fruit sets and after harvest annually. Check trees for common disease and fruit pests in the spring and fall.

Harvesting and Storage

Harvest cherries, stem on fruit, when they are fully ripe. Avoid damaging the spur (point of attachment) or you may damage next year's cherry. Fruit will be dark red, black or yellow depending on the variety. The sugar content in the fruit rises in the last few days of ripening so it's worth the wait. Fresh cherries will keep up to a week in your refrigerator, up to three in slightly cooler (31º to 32ºF) temperatures.

Choosing Cherries!

From the small wild cherry thousands of years ago, man's enjoyment of cherries has developed and we now expect to eat sweet varieties whenever in season.

We have wild cherry trees popping up everywhere in our garden. Thanks to the birds spitting out the pips on their own doorstep! Tut!

In our case the system can work well because the birds stay up in the heights of the old wild cherry trees, and tend to ignore the garden cherry trees tucked away in the vegetable plot. That's the theory but it doesn't always work like that. -Best to net your trees as soon as they start to fruit!

Because cherries are sooo particular, many varieties have been developed to cope with different temperatures and viruses. When you buy your cherry trees check instructions for:

Pollination needs: as a rule sour cherries - the wilder varieties - are self-pollinating. Sweet cherries generally need cross-pollination and should be planted near a compatible variety.

Regional compatibilty: Double check the variety is suitable for your region. Growing cherry trees in extreme temperatures will require a very special variety.

Planting instructions: Growing cherry trees in your garden requires a little forethought. They are trees after all! There are a few dwarf varieties on the market and these may have specific planting instructions.

HOW TO GROW CHERRY FRUIT TREES

Cherries are one of the most popular fruits. Growing cherry trees is easy. Perhaps the biggest problem is keeping the birds from beating you to the harvest. It can also be difficult to bring in a bowl of cherries from the tree. They get eaten before they get to the house!

Cherry trees can grow 20-30 feet tall, or more. They produce a profusion of blooms in early spring. Cherry blossom time is a cause for celebration in many areas of the country. The fruit matures early in the season, and is harvested in late spring.

There are two basic types of cherries:

— *Sweet Cherries*—The ones you eat fresh from your tree. Home gardeners prefer sweet cherries that they can eat fresh.

— Sour Cherries—commonly used in baking. Most sour cherry trees are grown in orchards by farmers.

If you don't know the difference between a sweet and a sour cherry, taste a fresh one of each.....then you'll know for certain!!!

Wood from cherry trees is hard. It makes excellent lumber for furniture. It also is good as wood for your fireplace.

Propagation of Cherry Trees

Cherry trees are grown from seeds. The pit, or seed, in the center of each cherry is capable of producing a tree. If you discard some cherry pits into a flower garden, you may find young seedlings growing there the following year.

Nowadays, most homeowners and gardeners buy young cherry trees from a garden store. The trees are usually 4' to 8' tall, and gives you a few years head start towards your first crop.

You can also graft cherry branches onto other fruit trees.

Planting Cherry Trees

Transplant young cherry tree saplings into your yard or garden in the spring or summer. Most likely, the container of the tree you purchased, is made of a bio-degradable, heavy peat material. After planting, it will soften and break down in the moist soil, allowing the roots to easily penetrate it and grow. If the container is plastic or some other material, you will need to remove the roots from it during planting.

Dig a hole about three times as big and deeper than the root ball. Mix in plenty of rich compost into the hole, along with fresh, rich soil.

If it is in a peat container, cut slits in the container to speed up its decomposition, and to help the roots to emerge easier.

Plant the chery tree to a depth equal to where it is in the container. Do not plant it deeper.

Fill in around the roots with a mixture of compost and soil. Tamp down the soil around the tree.

If planted in a windy area, stake the tree so it will grow straight.

Water well and deeply. Repeat deep watering frequently, to help new roots to grow and spread.

How to Grow Cherry Trees

Once your cherry tree is established, it should grow well with little attention. Fruit tree fertiliser spikes for the first few years, will help young trees to get off to fast, healthy start.

Each spring, before buds open, apply a dormant oil fruit tree spray. This will kill a variety of insects. Apply a fruit tree spray frequently during the season, if you have problems with insects or disease. We strongly recommend you avoid spraying near harvest time. In all cases, follow the directions on the label for application and use.

Pruning Cherry Trees

Like other plants, pruning your cherry tree is healthy for it. This should be done by early spring, before the new year's growth begins.

Prune the tree yearly, to remove dead or unhealthy branches and limbs. Also prune in areas where growth is very bushy. This will increase sunlight and air penetration, to help the overall health and growth of the tree. You can also prune branches to maintain a shapely looking tree.

Annual pruning will help keep your tree smaller and more manageable, too.

Controlling Birds in Cherry Trees

If you don't work to keep birds from getting to your cherry tree, you stand to lose a serious portion of the harvest. Here are some methods that other home gardeners have used:

Pest netting - Smaller trees can be wrapped in netting. It is very effective until the tree grows so big that it becomes impractical to cover it.

Aluminum pie tins on a string - its an old favorite. But, it can be irritating to you and the neighbors.

Noise makers—Anything that is loud and sporadic will startle the birds.

Fake predators—Plastic and blow-up owls and snakes

Pruning the Cherry and Plum

The Morello varieties of the cherry form rather open, round-headed tops that need comparatively little pruning if a well-defined stem and top are established when first set in orchard. As a rule, in all parts of the country where they thrive they will bear well without pruning except in the cutting out of dead twigs as they are noticed without regard to season. If at any time quite large limbs are cut the wounds do not heal over as readily as the apple or pear and usually form rotten spots, starting the exudation of gum. But the young growth can be shortened without injury where it is desirable to thicken up the top in interior climates, where it is often an advantage.

The duke and sweet cherries are upright in habit, and many commercial growers head back the top of young trees in orchard. In two or three years the natural habit is changed from the spire-shape to that of a round-topped apple-tree form. This plan better shades the stem and large branches, and shades the fruit more perfectly from the sun. The after-pruning consists mainly in taking out dead wood as it appears.

Some of the best native plums on rich soil make so much growth during the early stages of bearing that much of the bearing wood is exposed to the sun and the fruit is spoiled by scalding. This open habit is corrected by cutting back the new growth two or three years in succession. When the heavy-bearing stage is reached, the long growths no longer appear and the only pruning required is taking out the dead wood.

Some of the Japan and European varieties are also thickened in the tops by cutting back half of the new growth in the dormant period for two or three years in succession, when the trees are coming into bearing, to protect the fruit and broaden the tops for shading the stems and main branches. But it is now conceded by growers that cutting back the tops does not increase bearing or size of fruit. Its only value as indicated above is to shelter the fruit and. better shade the stem and branches of some varieties.

Crown-grafting Pear, Plum, and Cherry

These fruits are not as certain to unite as the apple in grafting unless additional care is taken. The common plan of indoor grafting is by wedge and side-uniting, as shown in Fig. 2. This plan gives uniting cell-surface on both sides of the scion. The grafts are inserted in the crown and only one graft is made from each seedling. They are tied and waxed, as shown in the figure, with the liquid grafting-wax. In making the wedge it is desirable to cut both sides the same thickness. If the slope is made thickest on the outer edge the pressure at the bark and cinbium layer is not favorable to the union. In making a crue side cut in the crown of the stock it must be held in some way firmly. A barrel-stave fastened at the back side of the cutting-bench with a wire passing down to a foot-rest, as shown in Fig. 45, answers the purpose well, bat at the point where the stave rests in the stock it must bo padded to prevent injury when pressed down firmly.

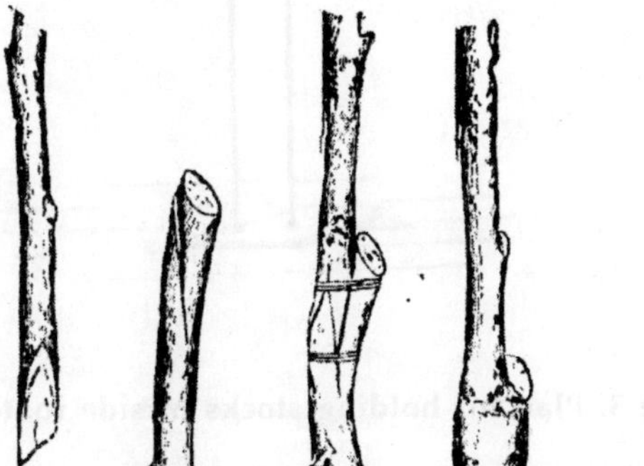

Figure 2. Side-grafting plum, cherry, and pear.

As the grafts are waxed they are rolled in sand to prevent sticking together. In sections where root-killing occurs the use of a long scion set down to the top bud in nursery is desirable where stocks are not entirely hardy. As the scion is inserted at the crown, and the saving of the fibrous part of the root is desirable, the use of the long scion gives a graft twelve or more inches in length. These long grafts need a deeper box in packing. However well cherry and plum grafts are made an even stand in nursery is unusual if the packed boxes are not stored in a cool cave. If the buds start prior to planting they usually fail to grow.

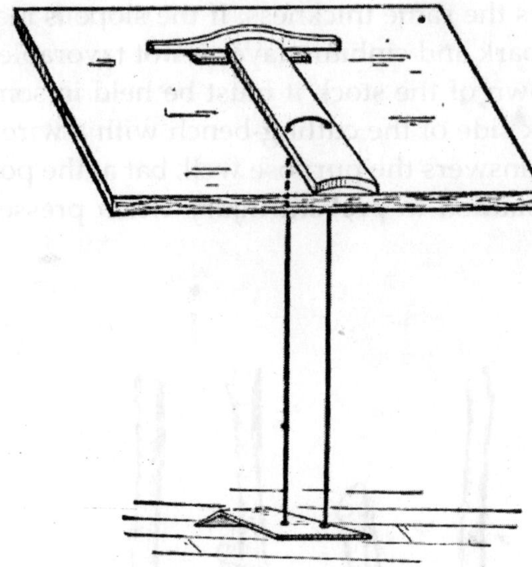

Figure 3. Plan for holding stocks in side root-grafting.

Top-grafting Cherry and Plum

Comparatively little has been done as yet in top-working the stone fruits. But Western experience with the cherry leads to the belief that it will pay commercially to top-work the cherry and plum on vigorous deep-rooting stocks. In Dupage County, Illinois, over forty years ago James Wakeman top-worked Early Eichmond on stocks now known as American Morello. The history of this variety is not known. It is worthless for fruit where better varieties can be grown, but it has remarkable vigor and hardiness and has been scattered by means of sprouts over the whole Northwest. The large commercial orchards top-worked on this stock have borne heavily and regularly and have outlived two or three generations of Early Eichmond nursery-grown trees on mahalob roots.

The same favorable experience has followed the use of this hardy stock in many parts of the West. At first the trees sprout but, when they come into heavy bearing the sprouting mainly ceases to give trouble. The wild red cherry (Prunus Pennsylvanicà) has also been used successfully as a stock. All varieties worked on it have proven hardier, lived longer, and have borne more regularly and profusely than the same varieties root-grafted or budded.

With the plum the benefits of top-working have been less apparent, except in the way of top-grafting select native varieties on native plum seedlings. Where the Japan and all foreign varieties have been top-grafted on native stocks they have come into bearing very young, but the top soon outgrows the stock and breaks off with the heavy load of fruit. This does not seem to arise from weakness of the stock, but from the hardening of the outer bark below the grafts. By slitting and peeling off the bands of outer cuticle the season after the grafting is done we have secured an even growth of scion and stock, both with the cherry and plum. Contrary to usual belief the top-working of cherry and plum is as easy and certain as grafting the apple or pear if the work is done very early with dormant scions and where the stocks are perfectly dormant. The side-cut method is employed on relatively small stocks, but on older stocks the cleft-and-wedge plan gives as good results.

Top-working in the Cellar

If sprouts or seedlings of any of the orchard fruits are taken up late in autumn and stored in cellar with proper root-covering they can be top-grafted during winter indoors and set in orchard the next spring. In this system a single scion is inserted in the stock at proper height to form the top. After grafting the only care needed is root-protection and care in handling to prevent breaking off the scion. When set in orchard or nursery shoots will start below the scion. These are left until the scion has made some growth, when they are rubbed off. This method is useful to amateurs, but is rarely practised in nurseries except with some shrubs, and the grafting of the gooseberry on Ribes aureum stocks.

Propagation of the Cherry

The selection of stocks and propagation by budding and grafting are given in Chapter VII (Propagation By Budding And Grafting) and the transplanting, spacing, and after care in Chapters IX and X. The most important consideration in prairie propagation of the cherrry is that of selection of a hardy stock on which to bud or graft the hardiest known varieties. This is specially true north of the 42d parallel. As stated on prior page all varieties of the cherry take well by budding on our native bird-cherry. At the experimental farm at Ottawa, Canada, during recent years, about the only cherry-trees that escaped root-killing were those on bird cherry-roots, and the same has been true in the prairie States. As Professor Bailey says, the bird cherry is quite a sprouter in its native localities. But when budded with the cultivated cherry we have seen but few sprouts on trees set twelve years in orchard. The draw on food-supply of the larger top and crops of larger fruit varieties seem to leave no reserve supply of nutriment for throwing up sprouts. But this only applies to the sections of our broad country where root-killing is frequent in open winters. Where the mazzard and mahaleb stocks are hardy they will continue to be the commercial stocks.

No Market Supply of the Cherry

As we now have varieties of good quality for dessert use or canning in about all parts of the Union it is a peculiar fact that our markets are not supplied except at a few local points, and in a small way in crates put up on the west coast for dessert use at prices beyond the reach of the masses. As Professor Bailey says: "The cherry is not cultivated as a leading industry east of the Rocky Mountains except in western New York, where the sour varieties are grown for canning. The sweet cherry is confined mostly to dooryard and fence-corner plantings. Sour kinds are found in orchard blocks in New York, New Jersey, Pennsylvania, Ohio, Michigan, Indiana, Illinois, Kansas, and Nebraska."

It is true that relatively small orchards are found in all these States. Yet it is also true that such orchards are far apart as a rule and the cherry is rarely found in market for family use, as is the case with all other orchard fruits. There is no reason for this, except that as yet the commercial planting of the cherry is neglected, especially west of the Great Lakes. Even in the small city of Ames, Iowa, fifty varieties of fine cherries were picked from healthy trees for exhibit at the Trans-Mississippi and International Exposition at Omaha. Yet the local demand for family canning is so poorly supplied that cherries are engaged a year in advance by many families. But this was true of the blackberry supply five years ago, while at this time every grocery in the Northwest has its blackberry supply following that of the strawberry.

On hardy roots the identical varieties of the cherry used for roadside trees in Germany, north Silesia, Switzerland, and south Russia will mainly thrive as well in this country, and the writer repeats the desire of Charles Downing, who said nearly forty years ago: "We wish we could induce the planting of avenues of this fine-growing fruit tree in our country neighborhoods, as is the beautiful custom in Germany, affording ornament and a grateful shade and refreshment to the traveller at the same moment."

REFERENCES

Cherry Production National Agricultural Statistics Service, USDA, Retrieved on August 19, 2008.

Tall JM, Seeram NP, Zhao C, Nair MG, Meyer RA, Raja SN (2004). "Tart cherry anthocyanins suppress inflammation-induced pain behavior in rat". *Behav. Brain Res.* 153 (1): 181?"8. doi:10.1016/j.bbr.2003.11.011. PMID 15219719.

"Tart Cherries May Reduce Heart/Diabetes Risk Factors". Newswise, Retrieved on July 7, 2008.

"FAOSTAT: ProdSTAT: Crops". Food and Agriculture Organization. 2007. http://faostat.fao.org/site/567/DesktopDefault.aspx?PageID=567. Retrieved on 07-02-2009.

5

Pear

The pear is an edible pomaceous fruit produced by a tree of genus *Pyrus*. The pear is classified within Maloideae, a subfamily within Rosaceae. The apple (*Malus ×domestica*), which it resembles in floral structure, is also a member of this subfamily.

The English word pear is probably from Common West Germanic *pera*, probably a loanword of Vulgar Latin *pira*, the plural of *pirum*, akin to Greek api(r)os, which is likely of Semitic origin. The place name Perry can indicate the historical presence of pear trees. The term "pyriform" is sometimes used to describe something which is "pear-shaped".

The cultivation of the pear in cool temperate climates extends to the remotest antiquity. Many traces of it have been found in the Swiss lake-dwellings. The word "pear" or its equivalent occurs in all the Celtic languages, while in Slavonic and other dialects different appellations, but still referring to the same thing, are found-a diversity and multiplicity of nomenclature which led Alphonse de Candolle to infer a very ancient cultivation of the tree from the shores of the Caspian to those of the Atlantic.

Pears grow in the sublime orchard of Alcinous, in *Odyssey* vii: "Therein grow trees, tall and luxuriant, pears and pomegranates and apple-trees with their bright fruit, and sweet figs, and luxuriant olives. Of these the fruit perishes not nor fails in winter or in summer, but lasts throughout the year."

The pear was cultivated also by the Romans, who did not eat them raw: Pliny's Natural History recommended stewing them with honey and noted three dozen varieties. The Roman cookbook attributed to Apicius, De re coquinaria, has a recipe for a spiced stewed-pear patina, or soufflé.

A certain race of pears, with white down on the under surface of their leaves, is supposed to have originated from *P. nivalis*, and their fruit is chiefly used in France in the manufacture of perry. Other small-fruited pears, distinguished by their early ripening and apple-like fruit, may be referred to *P. cordata*, a species found wild in western France, and in Devonshire and Cornwall. Pears have been cultivated in China for approximately 3000 years. The genus is thought to have originated in present-day western China in the foothills of the Tian Shan, a mountain range of Central Asia, and to have spread to the north and south along mountain chains, evolving into a diverse group of over 20 widely recognised primary species.

Figure 1. Pear Fruits

The enormous number of varieties of the cultivated European pear (*Pyrus communis* subsp. *communis*), are without doubt derived from one or two wild subspecies (*P. communis* subsp. *pyraster* and *P. communis* subsp. *caucasica*), widely distributed throughout Europe, and sometimes forming part of the natural vegetation of the forests. In England, where an ancient pear tree gave its name to Pirio in Domesday, the pear is sometimes considered wild; there is always the doubt that it may not really be so, but the produce of some seed of a cultivated tree deposited by birds or otherwise, which has germinated as a wild-form spine-bearing tree. Court accounts of Henry III of England record pears shipped from Rochelle and presented to the King by the Sheriffs of London.

BOTANY OF PEAR

Pears are native to coastal and mildly temperate regions of the Old World, from western Europe and north Africa east right across Asia. They are medium sized trees, reaching 10-17 m tall, often with a tall, narrow crown; a few species are shrubby. The leaves are alternately arranged, simple, 2-12 cm long, glossy green on some species, densely silvery-hairy in some others; leaf shape varies from broad oval to narrow lanceolate. Most pears are deciduous, but one or two species in southeast Asia are evergreen. Most are cold-hardy, withstanding temperatures between -25 °C and -40 °C in winter, except for the evergreen species, which only tolerate temperatures down to about -15 °C.

The flowers are white, rarely tinted yellow or pink, 2-4 cm diameter, and have five petals. Like that of the related apple, the pear fruit is a pome, in most wild species 1-4 cm diameter, but in some cultivated forms up to 18 cm long and 8 cm broad; the shape varies in most species from oblate or globose, to the classic pyriform 'pear-shape' of the European Pear with an elongated basal portion and a bulbous end.

The fruit is composed of the receptacle or upper end of the flower-stalk (the so-called calyx tube) greatly dilated. Enclosed within its cellular flesh is the true fruit: five cartilaginous carpels, known colloquially as the "core". From the upper rim of the receptacle are given off the five sepals, the five petals, and the very numerous stamens.

The pear is very similar to the apple in cultivation, propagation and pollination.

Pears and apples cannot always be distinguished by the form of the fruit; some pears look very much like some apples. One major difference is that pears have "grit" - clusters of lignified cells. Pear trees and apple trees do have several visible differences. The pear and the apple are also related to the quince. There are about 30 primary species, major subspecies, and naturally occurring interspecific hybrids of pears.

CULTIVATION

The pear may be readily raised by sowing the pips of ordinary cultivated or of wilding kinds, these forming what are known as free or pear stocks, on which the choicer varieties are grafted for increase. For new varieties the flowers can be cross-bred to preserve or combine desirable traits. The fruit of the pear is produced on spurs, which appear on shoots more than one year old.

Three species account for the vast majority of edible fruit production, the European Pear *Pyrus communis subsp. communis* cultivated mainly in Europe and North America, the Chinese white pear (bai li) *Pyrus ×bretschneideri*, and the Nashi Pear *Pyrus pyrifolia*, both grown mainly in eastern Asia. There are thousands of cultivars of these three species. A species grown in western China, *P. sinkiangensis*, and *P. pashia*, grown in southern China and south Asia, are also produced to a lesser degree.

Other species are used as rootstocks for European and Asian pears and as ornamental trees. The Siberian Pear, *Pyrus ussuriensis* (which produces unpalatable fruit) has been crossed with *Pyrus communis* to breed hardier pear cultivars. The Bradford Pear (*Pyrus calleryana* 'Bradford') in particular has become widespread in North America and is used only as an ornamental tree. The Willow-leafed Pear (*Pyrus salicifolia*) is grown for its attractive slender, densely silvery-hairy leaves.

HARVEST

Summer and autumn pears are gathered before they are fully ripe, while they are still green, but snap off when lifted. If left to ripen and turn yellow on the tree, the sugars will turn to starch crystals and the pear will have a gritty texture inside. In the case of the 'Passe Crassane', long the favored winter pear in France, the crop should be gathered at three different times, the first a fortnight or more before it is ripe, the second a week or ten days after that, and the third when fully ripe. The first gathering will come into eating latest, and thus the season of the fruit may be considerably prolonged.

USES

Pears are consumed fresh, canned, as juice, and dried. The juice can also be used in jellies and jams, usually in combination with other fruits or berries. Fermented pear juice is called perry. Pears will ripen faster if placed next to bananas in a fruit bowl. They stay fresh for longer if kept in a fridge. Pear wood is one of the preferred materials in the manufacture of high-quality woodwind instruments and furniture. It is also used for wood carving, and as a firewood to produce aromatic smoke for smoking meat or tobacco.

Health Benefits

Pears are rich in Vitamin A, Vitamin C, E1, copper and potassium. Pears are the least allergenic of all fruits. Because of this, it is sometimes used as the first juice introduced to infants. Along with lamb and soya formula, pears form part of the strictest exclusion diet for allergy sufferers. Pears can be useful in treating inflammation of mucous membranes, colitis, chronic gallbladder disorders, arthritis, and gout. Pears can also be beneficial in lowering high blood pressure, controlling blood cholesterol levels, and increasing urine acidity. They are good for the lungs and the stomach. Most of the fiber is insoluble, making pears a good laxative. Th gritty fiber content may cut down on the number of cancerous colon polyps. Most of the vitamin C, as well as the dietary fiber, are contained within the skin of the fruit.

PEAR VARIETIES

Most of North American pears are grown in Oregon and Washington, and the harvest months listed here reflect that. You can find some variety of pear in season in North America from August through May (and even into June some years). Your best bet for finding local pears is at farmers markets, ask the grower when you can expect the harvest and how long it might last.

Anjou Pears

Anjou pears have a mild flavor and a firm texture, while still being sweet and juicy. They can be green or red (which are often labelled "Red Anjou"), but the color doesn't indicate any major flavor or texture difference. Anjous are great for eating out of hand or for cooking.

Bosc Pears

Bosc pears are crisp when raw and hold their shape beautifully when cooked. They have the best example of the soft, grainy texture associated with pears. They are easy to recognise by their brownish russet skin, classic pear shape, and heady pear aroma. Bosc pears are harvested September through April.

Comice Pears

Comice pears are perhaps the best pears for eating raw. They have a great fruity aroma and flavor and a slightly finer, less grainy texture than other varieties. They are slightly more rounded and apple-shaped than other pears. Comice pears are harvested September through February.

Asian Pear

The term Asian pear describes a large group of pear varieties having crisp, juicy fruit. When mature, the fruit are good to eat when harvested or for several months after picking if held in cold storage. The crisp texture of an Asian pear remains unchanged after picking or storage, unlike the flesh of European pears such as Bartlett or Comice. There are three types of Asian pear: round or flat fruit with green to yellow skin; round or flat fruit with bronze-colored skin and a light bronze-russet; and pear-shaped fruit with green or russet skin.

Pear, common name for about 20 species of trees of a genus in the rose family, and for their fruit. The common pear is native to Europe; the Chinese sand pear is native to the Orient. Both species are extensively cultivated for their fruit in cool, humid, temperate regions throughout the world.

Under cultivation, standard pear trees attain heights of up to 9 m (30 ft), with trunks 30 cm (12 in) or more in diameter. The leaves are oval and simple and, unlike those of the apple, smooth and glossy. The white flowers, which are borne in umbels, have five sepals, five petals, many stamens, and a single pistil.

The fruit is a pome, juicier than the apple, and varying from apple-shaped to teardrop-shaped. Among different varieties, the thin skin varies in color from light yellow and green through red and brown. The thick flesh varies in flavor among different varieties. In young, unripe common pears, and in young and mature Chinese sand pears, the flesh contains numerous gritty cells called stone cells.

Pears are gathered from the trees before they are completely ripe and are allowed to ripen in storage. Cold retards ripening, and heat speeds it. Pears are eaten fresh and canned.

Commercial pear production in the United States averages about 700,000 metric tons annually. The best North American pear-growing districts are in California, Washington, and Oregon and, to a lesser degree, in the northern United States from New England to the Great Lakes and in lower Canada. Pears are grown extensively in home orchards in the United States.

Most pear varieties may be grown in either standard or dwarf sizes. Dwarf pears are propagated by grafting a pear scion on a quince stock. Angoul?e, Elizabeth, Louis Bonne, and Deal pears are desirable dessert varieties usually cultivated as dwarfs. Anjou, Boussoc, and Tyson pears are about equally good in either standard or dwarf sizes. Bosc, Washington, and Dix pears are usually grown as dwarfs by a type of grafting called double working, in which stocks are grafted onto stocks that have previously been dwarfed by grafting on quince. Bartlett, Seckel, and Doyenn?pears are usually grown in standard sizes.

Basic nutritional facts: Pears contain about 16 percent carbohydrate and negligible amounts of fat and protein. They are good sources of the B-complex vitamins and also contain vitamin C; in addition, they contain small amounts of phosphorus and iodine.

Scientific classification: Pears belong to the family Rosaceae. The common pear is classified as Pyrus communis and the Chinese sand pear as Pyrus pyrifolia.

CHOOSING A PEAR TREE

Pear trees bear fruit for anywhere between 60 and 200 years so give some careful consideration to the variety of pear tree you choose to grow. The factors involved in choosing a pear tree are size (controlled by the rootstock), táste and size (the variety) and pollination. Each is discussed in detail below.

Rootstock

Pears are so bland in the shops because they are all the same couple of varieties. Chosen so that they travel well, these are not the delicious pears you can grow in your garden. We explain how to plant, prune, care and harvest from your garden. We do it step by step in easy to understand language.

First decide how big you want your tree to grow. The eventual size of a pear tree depends on the rootstock. Pears are "grafted" onto rootstock. This means that the lower trunk of the tree is from a different tree to that of the top part of the tree. The reason this is done is because a pear tree grown on its own trunk and roots would be too large for most gardens. The fruit would certainly be out of reach without a ladder.

There are two commonly available rootstocks, Quince A and Quince C. There are minor differences between the two but nothing hugely significant.

Quince A is the most common and will enable you to grow a pear tree at maturity somewhere between 3m to 6m (9ft to 19ft). Regular pruning can keep the tree to size you want within that range. Very fertile soils will produce larger and quicker growing trees compared to less fertile soils. Fruit will appear after 5 years.

Quince C is less common but still readily available. At maturity the tree will be 3m to 5m (9ft to 16ft), not a great size difference compared to Quince A. Pear trees on Quince C are slightly quicker to produce fruit but the difference will only be one year. Much depends on the pear tree variety. In general, fruit will appear after 4 to 5 years.

Factors to be considered include the site where the pear tree is to be grown. Some varieties tolerate colder and windy conditions better than others. Disease resistance varies considerably between different varieties. As for taste, some pears taste better than others or are juicier. All these factors need to be considered when growing a pear tree. Hopefully the table below will help you in that choice. Although the pollination group for each variety is shown below, going to the next page will show you an easy to understand table of compatible pollinators.

POLLINATION

The majority of pear trees require another suitable variety nearby to enable fertilisation to occur. A few varieties such as Conference are self-fertile but they also produce better crops when a matching variety is nearby.

The question of "what is nearby" then arises. Bees come to your rescue here because they will fly several miles for a good pollen source.

So if you live near an allotment or are in a dense urban area it is likely that other people will be growing pear trees and amongst these will be several matching pollinators. Unfortunately, pear trees are rarer than apple trees so this may not always be the case.

Different varieties of pears flower at slightly different times of the year but almost all overlap to some degree. The best way to ensure the best pollination rate is to plant at least two pear trees which flower at roughly the same time. Pick partners based on the groupings below. Early flowering pear varieties include:

— Beth

- Beurre Hardy
- Conference
- Invincible
- Louise Bonne of Jersey
- Packham's Triumph
- Sensation
- William's Bon Chretien

Late flowering pear varieties include:

- Concorde
- Doyenne du Comice
- Glou Morceau
- Onward

When you have chosen which variety of pear tree you want and the rootstock you need to decide if you will buy a bare-rooted tree or a potted one.

Cost and timing are the major factors to consider. Potted trees are more expensive than bare rooted trees, by a significant amount. Against that, bare-rooted trees can only be planted when they are fully dormant, normally between mid December and early March. Potted trees can be planted at any time of the year although even these are best planted in winter.

Our recommendation is to go for a bare rooted tree, ordered online and delivered to your door from a reputable supplier. There are many reputable suppliers and recommendation from a fellow gardener is the best way to select one.

HOW TO PLANT A PEAR TREE

Where to Plant

Pear trees originate from Mediterranean areas and they appreciate similar conditions. Avoid planting in frost pockets or areas where they will be exposed to strong winds.

The best soil is a well dug but firm loam soil. Pear trees are not picky about the soil acidity unless it is far off the normal. They much prefer a moist soil and will not do well in soil which regularly dries out, especially for the first few years.

How to Plant

Planting a pear tree is exactly the same as planting an apple tree. With pear trees the support can be removed after four or five years although it will not harm it if left in place. Plant bush pear trees about 3.5m (11ft) apart if you have room for more than one.

Pear trees should be pruned annually in winter when they are dormant.

How to Prune Bush Pear Trees

Not surprisingly pruning pear trees is similar to pruning apple trees. The principal aim is to form a wine goblet shape over the years with the centre of the tree uncluttered to let air circulate.

We will give a link to pruning apple trees lower down but it is important to understand that there are two main differences between pruning apple trees and pear trees. Pear trees bear their fruit closer to the main branches compared to apples. For this reason allow around 8 main branches to form rather than the four or five of an apple tree.

The second difference which affects how to prune pear trees is that the braches aren't as strong as those of apple trees and are liable to snap . To build up the strength of the branches, prune as for apple trees but continue the formative pruning for eight years rather than four. This means repeating the pruning for year four through years five to eight. This will build up the strength of the pear tree branches.

Thinning out Fruit

In mid June pear trees will naturally drop some fruit. When this has finished, around mid July thin out the fruits even more. As a guide, snip of fruits so that the remaining fruits are roughly 10cm (4in) apart. Harsh though this may seem it will greatly assist the remaining fruit to develop to a good size.

Harvesting Pears

Pears need to be harvested at the correct time to be enjoyed at their best. Very early pears can be eaten immediately they are picked from the tree. Look out for when the skin begins to change colour at the bottom of the pear. Early pears will need about a week in storage before they are at their best for eating. Pick them when the flesh begins to change colour around the stalk and store for a week before eating.

How to Care For Your Pear Tree

Aside from pruning, pear trees require little attention. They do not like to be short of water though. The best strategy is to apply a mulch in mid-Spring of well-rotted compost around but at least 15cm (6in) away from the main trunk.

Pear Midges are often a problem. The cure is rake the surface of the soil regularly in from late winter to early spring. This will expose and kill the midges before they get a chance to fly up into the tree.

How to Store Pears

Pears don't store very well but some varieties are OK if kept for two or three weeks. Store pears in a cool area (approximately 6°C / 43°F) which doesn't suffer from large temperature changes. Lay them on a slatted surface and don't have more than one layer. Only store perfect specimens and check them out every couple of days. If any show signs of deteriorating then remove them immediately. A moist but not damp atmosphere is best.

PEST AND DISEASES

Pears are reasonably healthy fruit trees but do occasionally suffer from pest or disease the main ones are outlined below.

Aphids - these may attack new growth especially in early summer. The aphids can often be seen. They cause new shoots to curl and there is often a sticky fluid over new growth. Click here for more information and how to treat this pest.

Scab - this fungal disease can be identified by brown scab like marks on the fruit and often black marks on the leaves. Click here for details on how to treat this disease.

Canker - the symptoms are deformed patches of bark which becomes cracked and bends inwards. Click here for details on how to treat canker.

Caterpillars - these seem to find the leaves of pear trees particularly tasty. The best treatment is to simply pick them off by hand as soon as they are noticed. If you notice any moth cocoons them pick those off as well because they are housing caterpillars.

Pear Midge - the signs of pear midge are young fruit which does not develop properly and becomes black before falling to the ground. Destroy all the infected fruit. In spring cultivate the soil around the tree to expose the over-wintering midges to the cold and wild life. A mulch of black plastic over the soil around the tree will also prevent the midges entering the tree in Spring.

Fireblight - the signs are new growth turning brown and a sticky liquid appearing around affected areas in spring. Seek advice from your local garden centre if you suspect fireblight because this disease should be notified to the agricultural authorities.

PEAR PRODUCTION IN THE HOME GARDEN

Pears are long-lived, attractive trees that work well in Arkansas landscapes and produce fruit for home owners. In general, pears do not have as many pests as apples and are commonly grown by homeowners without pesticides. Selected varieties produce good fruit and present few management problems. The three basic types of pears grown in the United States are European or French pears, Oriental hybrids and Asian pears. The common pears include such popular varieties as 'Bartlett,' 'Bosc' and 'D'Anjou'. These and other common European varieties are especially noted for their excellent fresh eating quality. However, the susceptibility of European pears to fire blight limits their use.

Oriental hybrids include such well-known varieties as 'Orient' and 'Kieffer'. This type is well adapted to much of Arkansas and accounts for most of the state's pears. Oriental hybrid pears produce russeted fruit that is firm to hard and has a high number of grit cells. These varieties are popular for preserves, for pickling or in jams, but most varieties are also satisfactory for fresh use if properly ripened.

The Asian pear, often termed "apple-pear,' is a third type that is gaining increased attention because of its high-quality fruit. Asian pears are relatively new to Arkansas. This crop seems to do well in northern Arkansas. The southern extent of their adaptation in the state has not been determined.

Site and Soil Requirements

Climatically, pears are adapted to all areas of Arkansas. Pears often have severe problems with fire blight, and gardeners should only plant varieties with high blight tolerance. It is possible to lessen the effects of fire blight with judicious sprays, but this is difficult for homeowners to do and is not encouraged.

Good drainage is an important soil requirement, although pears are more tolerant of poorly drained soil than most other fruit trees. Sandy soils are best, but garden trees can be grown in clay or heavy loam soils in most parts of Arkansas.

Plentiful sunlight is a key factor for maximum fruit production. Choose an area of the yard in full sun or nearly full sun. Morning sunshine is particularly important for early drying of dew, which will reduce the incidence of disease.

Pears bloom early and blossoms are subject to spring freeze damage, which occurs most often on pears planted in low areas (valleys, along streams, etc.).

Plant pear trees in the winter or early spring while they are dormant. When fruit trees arrive from the nursery, inspect them for damage and general condition. Do not accept trees if roots are not moist. Soak the roots in water for 30 minutes to an hour before planting.

If soil at the planting site is compacted, thoroughly work the soil with a shovel or tiller. A soil test should be done to determine the soil pH and nutrient needs. A soil pH of 6.0 to 6.5 is optimum for pears, but trees usually do well in soils from pH 5.0 to 7.5.

Dig a planting hole large enough to spread the root system in a natural position. Larger holes filled with topsoil are of no benefit unless the soil at the planting site is extremely poor (rocky, calcareous, etc.). Do not add fertilizer to the planting hole.

Trim off broken or damaged root parts before planting. Set the plants at the same depth at which they were growing in the nursery. Work soil in and around the roots, firming to eliminate air pockets as the hole is filled. Do not leave a depression around the tree. Water the tree thoroughly and check for air pockets. If the tree settles, gently lift it to the proper planting depth.

Cut off the newly planted tree at 24 to 30 inches and remove all side branches. This is necessary to compensate for roots lost when the tree was dug at the nursery.

Training and Pruning

Pruning a young tree controls its shape by developing a strong, well-balanced framework of scaffold branches. Remove or cut back unwanted branches early to avoid the necessity of large cuts in later years. The preferred method of training pear trees is described in Figures 1 through 6. The multiple leader system described in the figures offers several advantages over trees trained to a single trunk. The multiple leader system has more, but shorter, side limbs; there is more fruiting wood in the tree's upper portion in the early years; and there is no need to use spreaders to make trees grow wider. Also, in cases of severe fire blight damage, multiple leaders offer more chances of escape from serious injury than trees with a single leader.

Do major pruning in late winter; prune sparingly in summer. Remove suckers that grow from the base of the trunk as soon as they are noted in the summer. Suckers from the Calleryana pear rootstock are thorny and have leaves that are distinctly different from others in the tree. If not pruned, rootstock suckers often grow to become a significant part of the tree. Calleryana suckers bear tiny, worthless fruit. On older bearing trees, continue to prune as shown in Figure 6. Cut back the leaders by approximately 24 inches each winter. Thin crowded shoots as needed to allow light penetration into the tree. If fire blight becomes a serious problem, prune sparingly, since the vigorous shoots stimulated by pruning cuts are usually more susceptible to fire blight.

Fertilisation

Fertilisation of pear trees should be based on soil analysis results to avoid problems with nutrient deficiencies or toxicities, but if these results are not available, as a general rule for newly planted trees when growth begins, apply ½ cup of balanced fertiliser (13-13-13 or equivalent) in a 2-foot circle around the tree. Keep fertiliser at least 6 inches from the tree trunk to avoid fertiliser burn. Each spring after growth starts, apply ½ cup of 13-13-13 (or equivalent) per year of age through the fourth year. Continue to apply about 2 cups per tree each spring.

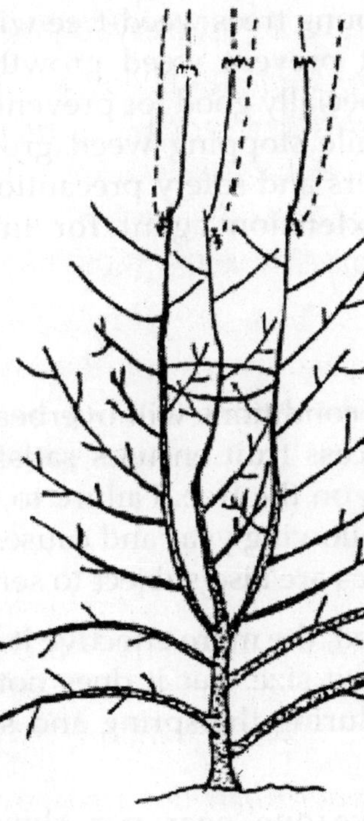

Figure 2. Cut back leaders and remove or cut back side shoots as described for the third winter.

If fire blight is a problem, discontinue fertiliser applications. If new growth is less than 6 inches per year, increase the amount of fertiliser. Mature trees growing in well-fertilised lawns generally receive adequate nutrition through the lawn fertilisation. Vigorous shoots are more vulnerable to fire blight, so use little or no fertiliser if blight is a problem. Use ammonium sulfate (21-0-0) instead of balanced fertiliser on highly alkaline soil (pH above 7.5) to avoid phosphorus-induced iron deficiency on highly acid soils.

Irrigation

In most sections of Arkansas, supplemental water is required for optimal tree growth and fruiting. Water young trees at least weekly. Mature pear trees are drought tolerant, but growth and fruiting are better if they are watered weekly or biweekly. Be sure irrigations are always sufficient to thoroughly soak the soil several inches deep.

Weed Control

Weed competition can result in death or poor growth of young trees. Keep an area at least the width of the canopy of young trees weed-free with a hoe, with plastic (or other types of mulching materials that prevent weed growth) or with chemicals. Woven polypropylene ground cover is especially good for preventing weed growth. It is durable and allows water penetration while stopping weed growth. Only applicators with a thorough knowledge of the dangers and safety precautions should use chemical weed killers. Consult your county Extension agent for information on weed control applications.

Fruit Thinning

Pear trees grown under favorable conditions will overbear, resulting in small fruit and often broken limbs. Removing excess fruit ensures satisfactory development of color, shape and size of pears remaining on the tree. Failure to remove excess fruit decreases formation of flower buds for the following year and causes trees to produce a good crop every other year. Overcropped trees are also subject to serious limb breakage problems.

The earlier thinning is completed, the more effective it is in achieving desired results. Midsummer thinning improves fruit size, but it does not aid formation of next year's flower buds, which are initiated during the spring and summer following full bloom. Thin fruit before this period.

Remove fruit by hand. Leave one pear per cluster, and space the clusters approximately every 6 inches. Start at one end of a branch and systematically remove fruit. To remove fruit without damaging other pears on the spur, hold the stem between the thumb and forefinger and push the fruit from the stem with the other fingers. This method removes the pears but leaves the stem attached to the spur.

REFERENCES

Galleta, Gene J., and David G. Himelrick, (eds.) 1990. *Small Fruit Management*. Englewood Cliffs, N.J.: Prentice Hall.

Mauseth, James D. 2003. *Botany: an introduction to plant biology*. Boston: Jones and Bartlett Publishers.

Nagy, Steven, and Philip E. Shaw. 1980. *Tropical and Subtropical Fruits:Composition, Properties, and Uses.* Westport, Conn.: Avi.

6

Peach

The peach (*Prunus persica*) is known as a species of Prunus native to China that bears an edible juicy fruit also called a peach. It is a deciduous tree growing to 5-10 m tall, belonging to the subfamily Prunoideae of the family Rosaceae. It is classified with the almond in the subgenus *Amygdalus* within the genus *Prunus*, distinguished from the other subgenera by the corrugated seed shell.

The leaves are lanceolate, 7-15 cm long (3-6 in), 2-3 cm broad, pinnately veined. The flowers are produced in early spring before the leaves; they are solitary or paired, 2.5-3 cm diameter, pink, with five petals. The fruit has yellow or whitish flesh, a delicate aroma, and a skin that is either velvety (peaches) or smooth (nectarines) in different cultivars. The flesh is very delicate and easily bruised in some cultivars, but is fairly firm in some commercial varieties, especially when green. The single, large seed is red-brown, oval shaped, approximately 1.3-2 cm long, and is a wood-like husk. Peaches, along with cherries, plums and apricots, are stone fruits. The tree is small, and up to 15 ft tall.

The scientific name persica, along with the word "peach" itself and its cognates in many European languages, derives from an early European belief that peaches were native to Persia. The modern botanical consensus is that they originate in China, and were introduced to Persia and the Mediterranean region along the Silk Road before Christian times. Cultivated peaches are divided into clingstones and freestones, depending on whether the flesh sticks to the stone or not; both can have either white or yellow flesh. Peaches with white flesh typically are very sweet with little acidity, while yellow-fleshed peaches typically have an acidic tang coupled with sweetness, though this also varies greatly. Both colours often have some red on their skin. Low-acid white-fleshed peaches are the most popular kinds in China, Japan, and neighbouring Asian countries, while Europeans and North Americans have historically favoured the acidic, yellow-fleshed kinds, though some prefer the other fruits.

Figure 1. peach fruits

Although its botanical name Prunus persica suggests the peach is native to Persia, it actually originated in China where it has been cultivated since the early days of Chinese culture. Peaches were mentioned in Chinese writings as far back as the tenth century BC and were a favored fruit of emperors.

Its English name derives from the Latin plural of persicum malum, meaning Persian apple. In Middle English, it melded into peche, much closer to what we call it today.

The Persians brought the peach from China and passed it on to the Romans. The peach was brought to America by Spanish explorers in the sixteenth century and eventually made it to England and France in the seventeenth century, where it was a prized albeit rare treat.

A horticultist by the name of George Minifie supposedly brought the first peaches from England to the USA in the early seventeenth century, planting them at his Estate of Buckland in Virginia.

Various American Indian tribes are credited with spreading the peach tree across the United States, taking seeds along with them and planting as they roved the country.

Although Thomas Jefferson had peach trees at Monticello, United States farmers did not begin commercial production until the nineteenth century in Maryland, Delaware, Georgia and finally Virginia. California grows 65% of peaches grown for commercial production in the United States, but the northern states, Colorado, Michigan, and Washington also grow a significant amount. Italy, China and Greece are major producers of peaches outside of the United States.

Peaches are known in China, Japan, Korea, Laos, and Vietnam not only as a popular fruit but for the many folktales and traditions associated with it. Momotaro, one of Japan's most noble and semi-historical heroes, was born from within an enormous peach floating down a stream. Momotaro or "Peach Boy" went on to fight evil oni and face many adventures. Peach flowers are admired by the Japanese but not as much as the sakura (cherry).

In China, the peach was said to be consumed by the immortals due to its mystic virtue of conferring longevity on all who ate them. The divinity Yu Huang, also called the Jade Emperor, and his mother called Xi Wangmu also known as Queen Mother of the West, ensured the gods' everlasting existence by feeding them the peaches of immortality. The immortals residing in the palace of Xi Wangmu were said to celebrate an extravagant banquet called the *Pantao Hui* or "The Feast of Peaches". The immortals waited six thousand years before gathering for this magnificent feast; the peach tree put forth leaves once every thousand years and it required another three thousand years for the fruit to ripen. Ivory statues depicting Xi Wangmu's attendants often held three peaches.

The peach often plays an important part in Chinese tradition and is symbolic of long life. One example is in the peach-gathering story of Zhang Daoling, who many say is the true founder of Taoism. Elder Zhang Guo, one of the Chinese Eight Immortals, is often depicted carrying a Peach of Immortality. The peach blossoms are highly prized in Chinese culture.

Due to its delicious taste and soft texture, in ancient China "peach" was also a slang word for "young bride", and it has remained in many cultures as a way to define young women or a general state of goodness.

A peach tree is also the context in which Kim Tr?ng and Thuý Ki?u fell in love in The Tale of Kieu. And in Vietnam, the blossoming peach flower is the signal of spring. Finally, peach bonsai trees are used as decoration during Vietnamese New Year - T?t in Northern Vietnam.

CULTIVATION

Peach plants grow very well in a fairly limited range, since they have a chilling requirement that tropical areas cannot satisfy, and they are not very cold-hardy. The trees themselves can usually tolerate temperatures to around -26 °C to -30 °C, although the following season's flower buds are usually killed at these temperatures, leading to no crop that summer. Flower bud kill begins to occur between -15 °C and -25 °C depending on the cultivar (some are more cold-tolerant than others) and the timing of the cold, with the buds becoming less cold tolerant in late winter. Certain cultivars are more tender and others can tolerate a few degrees colder. In addition, a lot of summer heat is required to mature the crop, with mean temperatures of the hottest month between 20 °C and 30 °C. Another problematic issue in many peach-growing areas is spring frost. The trees tend to flower fairly early in spring. The blooms often can be damaged or killed by freezes; typically, if temperatures drop below about -4 °C, most flowers will be killed. However, if the flowers are not fully open, they can tolerate a couple of degrees colder.

Important historical peach-producing areas are China, Iran, France, and the Mediterranean countries like Italy, Spain and Greece. More recently, the United States,

Canada, and Australia have also become important; peach growing in the Niagara Peninsula of Ontario, Canada, was formerly intensive but ended in 2008 when the last fruit cannery in Canada was closed by the proprietors. Oceanic climate areas like the Pacific Northwest and coastline of North Western Europe are generally not satisfactory for peach-growing due to inadequate summer heat, though they are sometimes grown trained against south-facing walls to catch extra heat from the sun. Trees grown in a sheltered and south-facing position in the southeast of England are capable of producing both flowers and a large crop of fruit.

For home gardeners, semi-dwarf (3 to 4 m) and dwarf (2 to 3 m) varieties have been developed by grafting desirable cultivars onto dwarfing rootstock. Fruit size is not affected. Another mutation is flowering peaches, selected for ornamental display rather than fruit production.

Depending on climate and cultivar, peach harvest can occur from late May into August; harvest from each tree lasts about a week.

Nectarines

The nectarine is a cultivar group of peach that has a smooth, fuzzless skin. Though fuzzy peaches and nectarines are commercially regarded as different fruits, with nectarines often erroneously believed to be a crossbreed between peaches and plums, or a "peach with a plum skin", they belong to the same species as peaches. Several genetic studies have concluded in fact that nectarines are created due to a recessive gene, whereas a fuzzy peach skin is dominant. Nectarines have arisen many times from peach trees, often as bud sports.

As with peaches, nectarines can be white or yellow, and clingstone or freestone. On average, nectarines are slightly smaller and sweeter than peaches, but with much overlap. The lack of skin fuzz can make nectarine skins appear more reddish than those of peaches, contributing to the fruit's plum-like appearance. The lack of down on nectarines' skin also means their skin is more easily bruised than peaches.

The history of the nectarine is unclear; the first recorded mention in English is from 1616, but they had probably been grown much earlier within the native range of the Peach in central and eastern Asia.

The trees are prone to a disease called leaf curl, which usually does not directly affect the fruit but does reduce the crop yield by partially defoliating the tree. The fruit is very susceptible to brown rot.

The developmental sequence of a nectarine over a 7½ month period, from bud formation in early winter to fruit ripening in midsummer. Most peach trees sold by nurseries are named cultivars grafted onto a suitable rootstock. It is also possible to grow a tree from either a peach or nectarine seed, but the fruit quality of the resulting tree will be very unpredictable.

Peaches should be located in full sun, and with good air flow. This allows cold air to flow away on frosty nights and keeps the area cool in summer. Peaches are best planted in early winter, as this allows time for the roots to establish and be able to sustain the new spring growth. When planting in rows, plant north-south.

For optimum growth, peach trees require a constant supply of water. This should be increased shortly before the harvest. The best tasting fruit is produced when the peach is watered throughout the season. Drip irrigation is ideal, at least one dripper per tree. Although it is better to use multiple drippers around the tree, this is not necessary. A quarter of the root being watered is sufficient.

Peaches have a high nutrient requirement, needing more nitrogen than most other fruit trees. An NPK fertiliser can be applied regularly, and an additional mulch of poultry manure in autumn soon after the harvest will benefit the tree. If the leaves of the peach are yellow or small, the tree needs more nitrogen. Blood meal and bone meal, 3-5 kg per mature tree, or calcium ammonium nitrate, 0.5-1 kg, are suitable fertilisers. This also applies if the tree is putting forth little growth.

If the full amount of peaches is left, they will be under-sized and lacking in sugar and flavour. In dry conditions, extra watering is important. The fruit should be thinned when they have reached 2 cm in diameter, usually about 2 months after flowering. Fresh fruit are best consumed on the day of picking, and do not keep well. They are best eaten when the fruit is slightly soft, having aroma, and heated by the sun.

Hard fruit can be softened at home at room temperature, though it may not get any sweeter. The process is facilitated by ethylene, similar to the ripening of bananas. Storage in a paper bag concentrates this gas and hastens the process. When the fruit is soft, it can be stored in the refrigerator for a few days to slow further changes.

In the August 13, 2004 edition of the Journal of Experimental Botany, researchers found that peaches stored around 40 °F (4 °C) became mealy and discolored.

After slicing, peaches exhibit enzymatic browning; ascorbic acid, usually in the form of lemon juice, is added to delay the process, and the flavors are complementary. Peaches also can be frozen, and can be blanched to inhibit enzymatic changes in texture. Peaches often are canned, both commercially and at home; added sugars and acids reduce the retort or pressure cooking time which reduces excessive softening.

A medium peach (75g), has 30 Cal, 7g of carbohydrate (6g sugars and 1g fiber), 1g of protein, 140mg of potassium, and 8% of the daily value (DV) for vitamin C. As with many other members of the rose family, peach seeds contain cyanogenetic glycosides, including amygdalin (note the subgenus designation: *Amygdalus*). These substances are capable of decomposing into a sugar molecule and hydrogen cyanide gas. While peach seeds are not the most toxic within the rose family, that dubious honor going to the bitter almond, large doses of these chemicals from any source are hazardous to human health.

PLANTING

Prepare the soil well in advance of planting to allow the soil to settle well - August is the ideal time. The planting area will be next to a wall, so dig a hole 1.8m (6ft) by 1m (3ft) to a depth of 75cm (2ft 6in). Dig the soil well, adding garden compost and/or other organic matter. Scatter two or three handfuls of bonemeal or other long-lasting fertiliser onto the soil and dig it in well.

Prior to planting the tree, attach wires to the wall as indicated in the diagram. Initially, the wires should cover an area of approximately 1.5m (4ft 6in) high by 2.5m (8ft) wide. This should be sufficient for a couple of years growth. Additional wiring can be attached in later years as the tree reaches maturity - see the diagram below.

When the peach tree is purchased (late September to early October is the best time) plant by spreading out the roots in the planting hole and filling in with soil to the same depth as the soil mark on the stem. The base of the stem should be 10cm (4in) out from the wall to allow for future growth. Firm the soil down. Lightly tie in any existing stems to the wire supports to allow for settlement of the soil.

TRAINING

In December to January of the first year (three months after planting the tree), when the new buds start to break, cut the tree back to roughly 60cm (2 ft) above ground. It is essential that at least two left and two right facing buds remain - one from each will form the principal structure of the tree. See the diagram to the right.

In the first summer, select the strongest left and right branches and remove all other shoots from the main stem. Cut back the central trunk back to 10cm (4in) above the highest branch. As the summer progresses, tie the two branches to the wires ensuring they are pointing upwards at a 45 degree angle. See the diagram to the left - note that at this stage of the year, the tree will be covered in foliage

In December to January of the second winter, cut the two side branches back to within 45cm (18in) of the main stem.

PRUNING

In the second summer (July), remove all growths from both of the main branches, leaving only three on each branch - two pointing upwards and one pointing downwards.

The branches should first be pruned around April time after the flowers have faded and the fruit is beginning to form.

Examine the branches which are bearing fruit and remove all new shoots on them with the exception of the following:

1. Do not remove the the shoot nearest the base of the branch — this should be allowed to develop and will bear fruit next year.
2. Keep three or four of the healthiest shoots which have young developing fruit.

The second phase of pruning is in December, when the foliage has died down and the structure of the tree is visible. Cut out all branches which have produced fruit this summer as far back as the shoot nearest the base of the branch (the 'DO NOT PRUNE' shoot shown in the above diagram). Tie in all stems to the wire supports.

Harvesting Peaches

Individual peaches need to be exposed to the sun if they are to ripen successfully. With this in mind, lightly prune any foliage which is clearly preventing light from reaching shaded peaches.

Knowing when the fruit is ripe can be judged from it's colour. If in doubt, gently press the fruit at the stalk end. If it 'gives' a little, the fruit is ripe - if it is hard, leave it for a week or so.

Ripe peaches do not keep for any length of time and they should be eaten within a day or so of harvesting.

DISEASES OF PEACH

Leaf Curl

The most important preventable pest for peach trees is peach leaf curl. The fungus 'Taphrina deformans', which is responsible for the disease, over-winters in cracks in the bark. When the young leaves expand, the fungus attacks them and results in severely deformed leaves. The flowers and fruit drop. Note that when the leaves have expanded to their full size, they can no longer be infected.

The cure for peach leaf curl is to spray the entire tree towards late January with Bordeaux Mixture - this is readily available in most garden centres. Follow the instructions carefully for best effect. Repeat the spraying in mid February . In this way, the fungus will be killed before the leaves expand.

Peach leaf curl is bad for for peach trees because it causes the leaves to become infected and fall off. This puts the tree under great stress and can sometimes kill the tree.

Often though the tree partially recovers and leaves begin to grow again in a month or so, not as good as before but they can grow. These new leaves will NOT become infected this season, whatever state the peach tree is in, so do not remove them because they are providing energy to a stressed tree.

The first thing to do is to remove all fruit from the tree. They will never develop fully after an attack of peach leaf curl so don't let the tree put any energy into them. Don't over-water but do provide water if the conditions become dry. Apply a nitrogen rich fertiliser around the base of the tree and water in well.

Peach leaf curl, caused by the fungal organism Taphrina deformans, is a common disorder in peach and nectarine orchards, especially during wet springs. Infected leaves become misshapen, deformed, and necrotic, resulting in premature defoliation with subsequent re-sprouting of new leaves. This kind of stress reduces fruit yield and predisposes the tree to pest attack. Some "tolerance" to peach leaf curl exists in the Redhaven series of peach cultivars, but none are truly resistant.

The infection period for leaf curl is when new leaves start emerging from buds in the spring. Spraying after the buds have opened is ineffective, because infection takes place as the young leaves emerge, and the fungus develops inside the leaf.

Accordingly, sprays must be applied during the trees' dormant period—after the leaves have fallen and before the first budswell in the spring. Many orchardists spray just prior to budswell during the months of February and March. Orchards with a history of severe peach leaf curl benefit from a double application: in the autumn at leaf fall and again in late winter or early spring just before budswell.

Fortunately for the organic grower, lime sulfur —one of the most effective fungicides for control of peach leaf curl—is allowed in certified organic production . Bordeaux and copper fungicides—also approved for certified organic programs —are effective as well, but not as effective as lime-sulfur.

Pscheidt and Wittig performed trials comparing Kocide™, lime-sulfur, several synthetic fungicides, and Maxi-Crop™ seav ed for leaf curl control. Lime-sulfur and one of the synthetics (ziram) were best, roughly twice as effective as Kocide. Seaweed sprays, despite positive anecdotal reports, were c mpletely ineffective.

Severe leaf curl infection can cause the tree to shed many of its leaves and to replace them with a second flush of growth. At this time the tree will benefit from a light feeding with a quickly-available soluble fertiliser such as compost tea or fish emulsion to help it recover.

There are various levels of resistance to leaf curl among varieties; however, because of the relative ease of controlling the disease, breeding for resistance has not been a priority. Redhaven, Candor, Clayton, and Frost are some of the cultivars with resistance to leaf curl, though none is immune. In contrast, Redskin and cultivars derived from it are susceptible.

Brown Rot

When peaches are grown under the warm humid conditions conducive to fungal diseases, it can be difficult to forego the use of fungicides. Brown rot (causal organisms: Monilinia fructicola and M. laxa) is foremost among fungal diseases, and peach producers struggle with it continually, as it affects both fruit yield and quality, and infests blossoms, twigs, and fruit in all stages. Brown rot is less prevalent west of the Rocky Mountains than in the East, but even in the West brown rot can be troublesome in seasonably wet or foggy microclimates.

Ideal conditions for infection arise during warm rainy periods (70–77°F is optimum). Brown rot occurs as blossom blight early in the growing season. Two to three weeks before harvest it infects the fruits as they soften and ripen, causing rot both at harvest and in storage— some of the infected fruit may not display symptoms until after harvest. Blossom blight during bloom is an indicator for extensive brown rot infections later in the season, although a wet year can produce heavy infections of brown rot from residual inoculum present in cankers and fruit, even without blossom blight.

In the East, control of brown rot is complicated not only by higher rainfall and humidity but also by increased levels of insect feeding, which spreads the inoculum and opens the fruit to infection. Moreover, the presence of alternate hosts such as wild plums and other wild Prunus species can further aggravate the situation. Under such conditions, commercial-scale organic production of peaches is currently extremely difficult.

Control of brown rot involves the integration of several tactics. Cultural practices and orchard sanitation are the first line of defense. Plantingsite selection and pruning are critical to providing sufficient air circulation within the canopy. Good air circulation through the tree facilitates rapid drying of the foliage and flowers after rain or overhead irrigation. Thinning branches to open the centre of the tree is a good practice— this can be done in July, as well as during the regular dormant-season pruning. Orchard sanitation practices include pruning and removal of infected twigs and cankers and disposal of dropped, culled, or mummified fruit. University of California researchers determined that excessive nitrogen fertilisation increases fruit susceptibility to brown rot. They also found that pre-harvest sprays of calcium reduced brown rot infection over non-sprayed trees but were not equal to fungicidal control.

Organic growers have traditionally relied on sulfur to control brown rot. The first application of sulfur should be done at the "pink" stage, just before the petals open. This should be repeated at seven-day intervals, especially if rain occurs, for a total of three applications. Two other applications should be made—one at petal drop, the other at sepal drop (usually about 10–14 days after petal drop). The crop is still susceptible to infection later in the season, but treatments during the early "critical" stage will reduce the amount of crop loss without leaving a sulfur residue at harvest. When the weather is hot and dry, the need to spray is not as great.

A promising organic control strategy for brown rot, according to Dr. Michael Glenn at USDA's Appalachian Fruit Research Station in Kearneysville, West Virginia, is to combine sulfur with Surround™ WP Crop Protectant. Derived from processed kaolin clay, Surround is an OMRI- (Organic Materials Review Institute) approved pest control product shown to control or suppress certain insects and diseases. The mechanism by which Surround suppresses powdery mildew, sooty blotch, fly speck, and fire blight (but not scab) in apples is thought to involve one or several of the following mechanisms: 1) physical separation of propagules from the plant surface, thus interfering with chemotaxic responses required for infection; 2) physical abrasion of the hyphae leading to ineffective infection; 3) physical dilution of nutrients at the plant surface that reduces the vigor and growth rate of the pathogen; and 4) removal of propagules from the plant surface through erosion of the particle film

Carl Rosato of Woodleaf Farm near Oroville, California, received funding from the Organic Farming Research Foundation (OFRF), the Kokaro Foundation, and the University of California, to test "natural" anti-fungal substances on his 3-acre peach orchard during the 1992–'94 growing seasons. The substances included compost tea, hydrogen peroxide, kelp sprays, grapefruit seed extract, rock dusts, a pink mucoid yeast, copper fungicides, vinegar, and combinations of these. In descending order, better control was obtained with : Algrow kelp mixed with basalt rock dust (55% marketable fruit); Algrow kelp alone (42%); compost tea + pink mucoid yeast (41%); hydrogen peroxide + pink mucoid yeast (40%); and so forth. Rosato's full research report, Peach Brown Rot Control, is available on the OFRF website.

In 2002, Rosato provided ATTRA with an update on his brown rot control strategy. First, it is helpful to understand that the part of California where his farm is located has a Mediterranean climate—a wet winter season alternating with a dry warm growing season—which is ideal for fruit production. Secondly, Rosato grows about 45 varieties of peaches selected for freshmarket quality as well as brown rot resistance.

For brown rot control, Rosato relies primarily on a spray mixture of micronised sulfur + rock dust (e.g., Azomite™). However, for a dynamic foliar spray that provides both nutritive and pest-control benefits, Rosato likes to blend a foliar "brew" for all pre-bloom, bloom, and postbloom sprays. A common tank mixture (per acre) may include: 6–8 lbs Azomite; 5–15 lbs micronized sulfur; 5 lbs soluble potassium sulfate; 1 lb Solubor™

(boron); 5 lbs kelp; and a yucca extract for a sticker. For the pre-bloom spray, he adds copper specifically for brown rot control. When peach twig borer is present, he adds B.t. (e.g., Dipel™). Bloom sprays begin at one-third bloom and proceed every 5–7 days all the way through petal fall, for a total of 3–4 sprays altogether. Postbloom sprays depend on the weather. When rain or humidity approaches, he religiously applies a brew spray as a prophylactic before weather arrives, again every 5–7 days depending on environmental conditions. Brown rot pressure decreases dramatically when it is hot and dry— around 85-90° F.

Rosato, who is also a soil consultant, emphasizes that nutrition is a fundamental aspect of orchard management. He follows the Albrecht approach to mineral balancing and soil management, paying attention to calcium-magnesium ratios, optimum micronutrient levels, and soil organic matter. For example, he applies 6 tons of compost per acre each year at the beginning of the growing season. He also feels the potassium and boron components of the brew mix contribute to brown rot control.

The Organic Farming Research Foundation also funded studies at Oregon State University, conducted by Hans Wittig and Dr. Jay Pscheidt between 1992 and 1995, that looked at anti-fungal properties of aerated compost tea, seaweed extracts, micronised sulfur, a yeast (Aureobasidium pullulans), and M-Pede™ insecticidal soap. Of these, insecticidal soap, sulfur, and a yeast + seaweed mixture were most effective in suppressing peach brown rot in the wet spring and arid summer conditions of western Oregon.

A new biofungicide, Serenade™, has demonstrated laboratory and field control of brown rot in California, but is not yet labeled for peaches. Serenade is a broad-spectrum preventive product registered for cherries, grapes, apples, pears, walnuts, hops, and vegetables. Diseases targeted include gray mold, powdery mildew, downy mildew, and sour rot. Serenade is OMRI approved for organic production, and organic peach growers are especially interested in the product.

A demonstration project in Fresno County, California, that was assessing the impact of composted yard trimmings in a commercial peach orchard found that composted green material not only compared favorably to other fertilising materials tested, but may also suppress brown rot in peaches.

Harvested fruit is also susceptible to brown rot infection. To prevent infections at harvest and during storage, peaches should be picked and handled with care to avoid punctures and skin abrasions on the fruit. Any damaged fruit should be discarded, since wounds facilitate entry of the fungus. Rapid cooling or hydrocooling to remove field heat prior to refrigeration at 0° to 3° C. will also help reduce infection.

Hot water treatments hold some promise for post-harvest control of brown rot. The hot water treatment takes two basic forms, mist or dip. In both cases the water is 50–52° C (approximately 122° F), but the mist treatment lasts approximately 15 minutes, while the dip lasts only 2–4 minutes. Chlorinated water can be used, and some additional chlorine added to help prevent post-harvest diseases, but note that organic certification standards in California permit a maximum of 10 ppm residual chlorine downstream of the product wash step.

Plum pox

In October 1999, the presence of plum pox, or Sharka Virus, was confirmed in Adams County, Pennsylvania—the first outbreak in North America. The disease has since been found in Cumberland County, PA, and in Canada. This has caused much concern among producers, because plum pox is an exceptionally destructive disease of stonefruit. Infected fruit are unmarketable because of spots and ring blemishes, and fruit may also drop prematurely. Plum pox is transmitted either by aphids or by grafting. The disease was not found in nursery stock, and an ongoing quarantine has apparently contained the outbreak of plum pox to a limited area, although testing continues.

Bacterial Spot

The tell-tale symptom of bacterial spot (caused by the bacterium Xanthomonas pruni) is small light-brown lesions on leaves. Eventually the affected tissue falls out, leaving a characteristic shotgun-hole appearance. Severe bacterial spot infections may cause premature defoliation and subsequent re-sprouting, similar to peach leaf curl. Bacterial spot on fruit occurs as sunken, dry lesions that eventually crack, opening the fruit for secondary infections and reducing fruit quality. Selecting disease-resistant cultivars is the principle means of controlling bacterial spot.

Fortunately for organic growers, copper fungicides—unique in that they also function as bactericides —are recommended for control of bacterial spot. The first spray should be applied before the tree leafs out in the spring; this timing often allows copper-based peach-leaf-curl sprays to double for bacterial spot treatment. The next period when infection pressure is heavy is petal fall and three weeks thereafter. Additional spray coverage may be necessary depending on varietal susceptibility and humid weather conditions. Since the occurrence and severity of bacterial spot depend on moisture, it is rarely a problem west of the Rocky Mountains, and in the East growers are able to rely on resistant varieties as the best line of defense.

Peach scab

Peach scab is caused by a fungus (*Cladosporium carpophilum*) that overwinters in twig lesions. Splashing rain spreads the fungal spores to young fruit and new shoots. Scab symptoms include small dark-green spots on the immature fruit. As the fruit matures, the spots enlarge and turn brown, and may cause the fruit skin to crack. Sulfur and most other fungicides that are applied for brown rot will also control peach scab. There are no resistant cultivars. Pruning to improve air circulation and reduce wetness in the tree can help manage the fungus and prevent twig infection.

Peach Mosaic Virus

First identified in the United States in 1931, peach mosaic virus has since appeared in a number of western states. Spread by grafting and the peach bud mite, the disease results in delayed and rosette foliation, low fruit production, and deformed fruit. The disease has largely been controlled in the United States through quarantines and destruction of infected trees. In areas where peach mosaic virus quarantines have occurred there may be restrictions on planting susceptible cultivars.

Fungicides

Concerns about the toxicity of fungicide and pesticide residues on food have prompted research that is yielding a new generation of safer and more effective fungicides. Some of these new fungicides are noteworthy because they meet guidelines for certified organic production. Other fungicides, though not permitted in organic programs, are based on new types of chemistry and biology that are seen as improvements over older fungicides.

New-generation fungicides approved for organic production are mostly microbial antagonists they consist of "good guy" fungi, yeasts, and bacteria that suppress "bad guy" plant pathogens. Among the microbials, one of the most promising introductions is Serenade™, a product of AgraQuest. The active ingredient in Serenade is a bacterium, Bacillus subtilis, QST 713 strain. This microbe inhibits germination of plant pathogen spores and restricts their growth once established through a number of biocontrol mechanims. Though Serenade is not currently labeled for peaches, future developments may result in a new tool for peach growers.

In addition to biofungicides, other disease control products that may be used in organic production include fungicidal soaps, copper and sulfur fungicides, peroxides, and botanicals (products derived from plants). Because botanicals break down quickly, they are less persistent in the environment. This makes them more challenging to use—the grower must time applications very carefully, and in some cases make them more frequently. This can be a significant cost consideration since botanicals are generally more expensive than synthetic pesticides. Also, botanicals—particularly when frequently applied —can harm beneficial insects. One of the most commonly used botanicals is Rotenone, derived from the roots of Southeast Asian and Central and South American plants. It is somewhat effective against a large number of insect pests, including the plum curculio and tarnished plant bug.

New classes of fungicides include strobilurins and phenylpyrroles, which were both initially discovered in nature (though fungicide formulations are synthesized analogs). Strobilurin, for example, was first isolated from wood-decaying European strobilurin mushrooms, while phenylpyrroles were first found in bacteria. The strobilurin fungicides Abound™ (azoxystrobin) and Flint™ (trifloxystrobin), and the phenylpyrrole fungicides Medallion™ (fludioxonil) and Scholar™ (fludioxonil), exhibit good activity against brown rot in peaches. Because these fungicides have low mammalian toxicity and short persistence in the environment, the EPA has classified them as reduced-risk pesticides.

The sterol inhibitors are the next class of new fungicides. They include Elite™ (tebuconazole), Indar™ (fenbuconazole), Nova™ (myclobutanil), and Orbit™ (propiconazole), all of which are registered for peaches and exhibit excellent brown rot control, yet boast very low mammalian toxicity. However, the nature of their mode of action predisposes them to development of resistance by the pathogen. One way to avoid this resistance is to rotate fungicide use among different chemical classes—for instance, alternate sprays of a strobilurin with a sterol inhibitor.

Insect Pests

The peach tree borer

The peach tree borer (Synanthedon exitiosa) and lesser peach tree borer (S. pictipes) can be major pests of peaches. Borers feed on the inner bark of trees, where they may kill the tree by girdling or cause the bark to peel away, exposing the tree to other pests and diseases. Other hosts for the borers include wild and cultivated cherry, plum, prune, nectarine, apricot, and certain ornamental shrubs of the genus Prunus. The adult peach tree borer is a clearwing moth, steel blue with yellow or orange markings. The moths are day fliers and can easily be mistaken for wasps.

These insects overwinter as larvae in burrows at the base of the host tree. Because the eggs are laid over a long period of time, the larvae vary greatly in size. Some are more than a half-inch in length, while others are very small, not more than an eighth of an inch long. The larvae pupate in the trunk of the tree, and usually begin to emerge as adults in June. Adult emergence and egg-laying occur from June through September, peaking during August.

The females are attracted to trees that have previously been damaged by borers, or to which some mechanical injury has occurred. Therefore, it is important to prevent damage to the tree trunk in order to minimise borer attack. Trees in poor vigor because of weed competition or drought stress also seem to be more susceptible to borer attack and damage.

Organic peach growers can use a variety of tactics to control this pest. Interior white latex paint can be painted or sprayed on the base of the trunks. This provides a physical barrier to the newly hatched larvae, inhibiting their entry into the trunk. The paint also fills cracks in the bark, where the larvae prefer to feed and the adult moths prefer to lay eggs. Because the paint is more a deterrent than a perfect control, some organic growers mix rotenone in with the paint; no data have been collected, however, to verify that the rotenone increases the paint's efficacy. It is easy to detect a tree that is infested with peach tree borers, since large amounts of gum exude from the damaged areas. The grower can Peach growers seeking to reduce pesticides can use a variety of tactics to control this pest. Interior white latex paint, painted or sprayed on the base of the trunks, provides a physical barrier, inhibiting newly hatched larvae from entering the trunk. The paint also fills cracks in the bark, the preferred site for oviposition and larval feeding. Because the paint is more a deterrent than a perfect control, some of these growers mix rotenone in with the paint; no data have been collected to verify that the rotenone increases the paint's efficacy. According to OMRI, use of latex paint is not allowed in organic production systems use this exudate to locate a larva, and then kill it by using a knife or flexible wire to probe it out of the trunk. The soil should be removed from around the base of the tree to a depth of three inches before starting this process, since larval damage also occurs under the soil line. This method of control is feasible for small plots but probably not practical in a commercial orchard.

The bacterium Bacillus thuringiensis (B.t.) can be used to control the larvae before they have entered the trunk. The products Dipel™, Thuricide™, and Javelin™ are formulations made from this organism. Because B.t. does not have a long residual effect, the trunk should be sprayed weekly with one of these materials during the period of peak moth flight, late July through August.

A biological control, the commercially available insect-parasitic nematode Steinernema carpocapsae, has also been used to successfully manage peachtree borers, when applied as a lower-trunk drench in warm spring or fall weather.

Peach twig borer

The peach twig borer (Anarsia lineatella) is only a minor pest in the eastern U.S., but it's a significant problem in Texas and the West. The larvae emerge in the spring and then bore into twigs and buds before pupating into dark gray moths. Later generations of larvae attack maturing fruit during the summer, entering fruit near the stem end and rendering it unfit for sale.

Treatment after borers have entered the tree is much less effective than treatment during the dormant or bloom period. Peach twig borers are usually not a problem in orchards that are sprayed each year at the delayed dormant period with lime-sulfur or with a 3% oil emulsion.

Two to three Bacillus thuringiensis sprays at bloom also appear to be effective against the twig borer.

The peach twig borer has many natural enemies and parasites, including the parasitic wasps Paralitomastix varicornis, Macrocentrus ancylivorus, Euderus cushmani, Hyperteles lividus, Erynnia species and Bracon gelechiae, as well as the grain mite Pyemotes ventricosus. The California gray ant, Formica arrata, can be beneficial when it preys on peach twig borer, but it unfortunately also protects aphids and scales.

Other predators of the peach twig borer include lacewings, ladybugs, and minute pirate bugs. These insects can be attracted to the orchard by habitat plantings, cover crops, and hedgerows. Mating disruption can also be effective if properly implemented.

REFERENCES

Billy, Simon. *Horticultural Farm Manager*. The Kerr Center for Sustainable Agriculture, Poteau, OK. Personal communication. May 1997.

Huxley, A., ed. (1992). *New RHS Dictionary of Gardening*. Macmillan ISBN 0-333-47494-5.

Szalay, L., Papp, J., & Szaóbo, Z. (2000). Evaluation of frost tolerance of peach varieties in artificial freezing tests. In: Geibel, M., Fischer, M., & Fischer, C. (eds.). Eucarpia symposium on Fruit Breeding and Genetics. *Acta Horticulturae* 538. Abstract.

7

Persimmon

A persimmon, known to the ancient Greeks as "the fruit of the gods" is the edible fruit of a number of species of trees of the genus *Diospyros* in the ebony wood family (Ebenaceae). The word *persimmon* is derived from *putchamin*, *pasiminan*, or *pessamin*, from Powhatan, an Algonquian language of the eastern United States, meaning "a dry fruit". Persimmons are generally light yellow-orange to dark red-orange in color, and depending on the species, vary in size from 1.5-9 cm (0.5-4 in) diameter, and may be spherical, acorn-, or pumpkin-shaped. The calyx often remains attached to the fruit after harvesting, but becomes easier to remove as it ripens. They are high in glucose, with a balanced protein profile, and possess various medicinal and chemical uses. While the persimmon fruit is not considered a "common berry" it is in fact a "true berry" by definition.

A Persimmon is a fruit of a number of species of trees of the genus Diospyros in the ebony wood family (Ebenaceae), and the edible fruit borne by them. The word persimmon is derived from putchamin, pasiminan, or pessamin, from Powhatan, an Algonquian language (related to Blackfoot, Cree and Mohican) of the eastern United States, meaning "a dry fruit".Persimmons are generally light yellow-orange to dark red-orange in color, and depending on the species, vary in size from 1.5-9 cm (0.5-4 in) diameter, and may be spherical, acorn-, or pumpkin-shaped. The calyx often remains attached to the fruit after harvesting, but becomes easier to remove as it ripens. They are high in glucose, with a balanced protein profile, and possess various medicinal and chemical uses.

The heart-shaped Hachiya is the most common variety of astringent persimmon. Astringent persimmons contain very high levels of soluble tannins and are unpalatable if eaten before softening. The astringency of tannins is removed through ripening by exposure to light over several days, or artificially with chemicals such as alcohol and carbon dioxide which change tannin into the insoluble form. This bletting process is sometimes jumpstarted by exposing the fruit to cold or frost which hastens cellular wall breakdown. These astringent persimmons can also be prepared for commercial purposes by drying.

Figure 1. Persimmon fruit

The oriental persimmon is native to China, where it has been cultivated for centuries and more than two thousand different cultivars exist. It spread to Korea and Japan many years ago where additional cultivars were developed. The plant was introduced to California in the mid 1800's.

Persimmons do best in areas that have moderate winters and relatively mild summers—suitable for growing in USDA Hardiness Zones 7 to 10. It can tolerate temperatures of 0° F when fully dormant. However, because of its low chilling requirement (less than 100 hours), it may break dormancy during early warm spells only to be damaged by spring frosts later. The leaves are killed by 26° F when growing. Trees do not produce well in the high summer heat of desert regions, which may also sunburn the bark.

The persimmon is a multitrunked or single-stemmed deciduous tree to 25 ft. high and at least as wide. It is a handsome ornamental with drooping leaves and branches that give it a languid, rather tropical appearance. The branches are somewhat brittle and can be damaged in high winds.

Foliage: Persimmon leaves are alternate, simple, ovate and up to 7 inches long and 4 inches wide. They are often pale, slightly yellowish green in youth, turning a dark, glossy green as they age. Under mild autumn conditions the leaves often turn dramatic shades of yellow, orange and red. Tea can also be made from fresh or dried leaves.

Flowers: The inconspicuous flowers surrounded by a green calyx tube are borne in the leaf axils of new growth from one-year old wood. Female flowers are single and cream-colored while the pink-tinged male flowers are typically borne in threes.

Commonly, 1 to 5 flowers per twig emerge as the new growth extends. Persimmon trees are usually either male or female, but some trees have both male and female flowers. On male plants, especially, occasional perfect flowers occur, producing an atypical fruit. A tree's sexual expression can vary from one year to the other. Many cultivars are parthenocarpic, although some climates require pollination for adequate production. When plants not needing pollination are pollinated, they will produce fruits with seeds and may be larger and have a different flavor and texture than do their seedless counterparts.

Persimmons can be classified into two general categories: those that bear astringent fruit until they are soft ripe and those that bear nonastringent fruits. Within each of these categories, there are cultivars whose fruits are influenced by pollination and cultivars whose fruits are unaffected by pollination. Actually, it is the seeds, not pollination per se, that influences the fruit. An astringent cultivar must be jelly soft before it is fit to eat, and such cultivars are best adapted to cooler regions where persimmons can be grown. The flesh color of pollination-constant astringent cultivars is not influenced by pollination. Pollination-variant astringent cultivars have dark flesh around the seeds when pollinated. A nonastringent persimmon can be eaten when it is crisp as an apple. These cultivars need hot summers, and the fruit might retain some astringency when grown in cooler regions. Pollination-constant nonastringent (PCNA) persimmons are always edible when still firm; pollination-variant nonastringent (PVNA) fruit are edible when firm only if they have been pollinated.

The shape of the fruit varies by cultivar from spherical to acorn to flattened or squarish. The color of the fruit varies from light yellow-orange to dark orange-red. The size can be as little as a few ounces to more than a pound. The entire fruit is edible except for the seed and calyx. Alternate bearing is common. This can be partially overcome by thinning the fruit or moderately pruning after a light-crop year. Astringency can also be removed by treating with carbon dioxide or alcohol. Freezing the fruit overnight and then thawing softens the fruit and also removes the astringency. Unharvested fruit remaining on the tree after leaf fall creates a very decorative effect. It is common for many immature fruit to drop from May to September.

The Black Persimmon or Black Sapote (*Diospyros digyna*) is native to Mexico. Its fruit has green skin and white flesh, which turns black when ripe.

The Mabolo or Velvet-apple (*Diospyros discolor*) is native to the Philippines. It is bright red when ripe.

The Japanese Persimmon or kaki (*Diospyros kaki*), "shizi" in Chinese, is the most widely cultivated species. These are sweet, slightly tangy fruits with a soft to occasionally fibrous texture. This species, native to China, is deciduous, with broad, stiff leaves.

Cultivation of the fruit extended first to other parts of east Asia, and was later introduced to California and southern Europe in the 1800s, and numerous cultivars have been selected. It is edible in its crisp firm state, but has its best flavor when allowed to rest and soften slightly after harvest. The Japanese cultivar 'Hachiya' is a widely grown cultivar. The fruit has a high tannin content which makes the immature fruit astringent and bitter. The tannin levels are reduced as the fruit matures. Persimmons like 'Hachiya' must be completely ripened before consumption. When ripe, this fruit comprises thick pulpy jelly encased in a waxy thin skinned shell. "Sharon Fruit" is the trade name for *D. kaki* fruit that has been artificially ripened with chemicals. It is also known as the "Asian mango".

The Date-plum (*Diospyros lotus*) is native to southwest Asia and southeast Europe. It was known to the ancient Greeks as "the fruit of the gods", i.e. *Dios pyros*, hence the scientific name of the genus. Its English name probably derives from Persian *Khormaloo* ÎÑãÇáæ literally "Date-Plum", referring to the taste of this fruit which is reminiscent of both plums and dates. This species is one candidate for the lotus mentioned in *the Odyssey*: it was so delicious that those who ate it forgot about returning home and wanted to stay and eat lotus with the lotus-eaters.

The American Persimmon (*Diospyros virginiana*) is native to the eastern United States and is higher in nutrients like vitamin C and calcium than the Japanese Persimmon.

There are many other species of persimmon that are inedible to humans, and thus have little or no commercial value for their fruit.

FRUIT

Commercially, there are generally two types of persimmon fruit: astringent and non-astringent.

The heart-shaped Hachiya is the most common variety of astringent persimmon. Astringent persimmons contain very high levels of soluble tannins and are unpalatable if eaten before softening. The astringency of tannins is removed through ripening by exposure to light over several days, or artificially with chemicals such as alcohol and carbon dioxide which change tannin into the insoluble form. This bletting process is sometimes jumpstarted by exposing the fruit to cold or frost which hastens cellular wall breakdown. These astringent persimmons can also be prepared for commercial purposes by drying.

The non-astringent persimmon is squat like a tomato and is most commonly sold as fuyu. Non-astringent persimmons are not actually free of tannins as the term suggests, but rather are far less astringent before ripening, and lose more of their tannic quality sooner. Non-astringent persimmons may be consumed when still very firm to very very soft.There is a third type, less commonly available, the pollination-variant non-astringent persimmons. When fully pollinated, the flesh of these fruit is brown inside -known as goma in Japan, and the fruit can be eaten firm. These varieties are highly sought after and can be found at specialty markets or farmers markets only. Tsurunoko, sold as "Chocolate persimmon" for its dark brown flesh, Maru, sold as "Cinnamon persimmon" for its spicy flavor, and Hyakume, sold as "Brown sugar" are the three best known.

Before ripening, persimmons usually have a "chalky" taste. They are generally only eaten when ripe and soft.

FRUIT PRODUCTION

The table below shows figures of persimmons for the world's top ten persimmon producing countries according to FAO statistics.

Table 1. Production figures in tonnes per year

Country	*1970*	*1990*	*1995*	*2000*	*2005*
China	457 341	640 230	985 803	1 615 797	1 837 000
Korea	30 310	95 758	194 585	287 847	250 000
Japan	342 700	285 700	254 100	278 800	230 000
Brazil	21 659	46 712	51 685	63 300	150 000
Italy	59 600	68 770	61 300	42 450	51 332
Israel	-	17 200	11 000	14 000	40 000
New Zealand	-	972	1600	1200	1300
Iran	25	925	1000	1000	1000
Australia	-	329	640	650	650
Mexico	-	275	274	450	450

USES

Culinary Uses

Persimmons are eaten fresh or dried, raw or cooked. When eaten fresh the peel is usually cut/peeled off and the fruit is often cut into quarters or eaten whole like an apple. The flesh ranges from firm to mushy and the texture is unique. The flesh is very sweet and when firm possesses an apple-like crunch. In China, Korea, Japan, and Vietnam after harvesting, 'Hachiya' persimmons are prepared using traditional hand-drying techniques, outdoors for two to three weeks. The fruit is then further dried by exposure to heat over several days before being shipped to market.

In Korea, dried persimmon fruits are used to make the traditional Korean spicy punch, *sujeonggwa,* while the matured, fermented fruit is used to make a persimmon vinegar called *gamsikcho,* which is alleged to have a variety of health benefits. The *hoshigaki* tradition traveled to California with Japanese American immigrants. A few farms still practice the art, which is being revived in part through the efforts of Slow Food USA, which describes the technique on its site and provides links to producers.

In Taiwan, fruits of astringent varieties are sealed in jars filled with lime water to get rid of bitterness. Slightly hardened in the process, they are sold under the name "crisp persimmon" or "water persimmon". Preparation time is dependent upon temperature (5 to 7 days at 25-28°C). In some areas of Manchuria and Korea, the dried leaves of the fruit are used for making tea. The Korean name for this tea is *ghamnip cha.*

The persimmon also figures prominently in American culinary tradition. It can be used in cookies, cakes, puddings, salads and as a topping for breakfast cereal. Persimmon pudding is a dessert using fresh persimmons. An annual persimmon festival, featuring a persimmon pudding contest, is held every September in Mitchell, Indiana. Persimmon pudding is a baked pudding that has the consistency of pumpkin pie but resembles a brownie and is almost always topped with whipped cream. Persimmons may be stored at room temperature (20°C) where they will continue to ripen. It is also a native plant in Brazil, South America, where it is referred to as the Caqui. In northern China, unripe persimmons are frozen outside during winter to speed up the ripening process.

Ethnomedical Uses

- In traditional Chinese medicine the fruit regulates ch'i
- The raw fruit is used to treat constipation and hemorrhoids, and to stop bleeding. As such, it is not a good idea to consume too many persimmons at once as they can induce diarrhea
- The cooked fruit is used to treat diarrhea and dysentery
- The apparent contradictory effect of t e raw and cooked fruit is due to its osmotic effect in the raw fruit sugar (causing diarrhea), and the high tannin content of the cooked fruit helping with diarrhea.

The fruits of some persimmon varieties contain the tannins catechin and gallocatechin, as well as the anti-tumor compounds betulinic acid and shibuol, although the latter may also cause gastrointestinal problems.

Unripened persimmons contain the soluble tannin shibuol, which, upon contact with a weak acid, polymerises in the stomach and forms a gluey coagulum that can affix with other stomach matter. The *Merck Manual of Diagnosis and Therapy* notes that consumption of persimmons has been known to cause bezoars that require surgery in over 90% of cases. More than 85% phytobezoars are caused by ingestion of unripened persimmons. Persimmon bezoars often occur in epidemics in regions where the fruit is grown.

Horses may develop a taste for the fruit growing on a tree in their pasture and overindulge also, making them quite ill. It is often advised that persimmons should not be eaten with crab meat, nor should they be eaten on an empty stomach.

Uses of Wood

Though persimmon trees belong to the same genus as ebony trees, persimmon tree wood has a limited use in the manufacture of objects requiring hard wood. It is hard, but easily cracks and somewhat difficult to process. Persimmon wood is used for paneling in traditional Korean and Japanese furniture.

In North America, the lightly colored, fine-grained wood of *D. virginiana* is used to manufacture billiard cues and shuttles. Persimmon wood was also heavily used in making the highest-quality heads of the golf clubs known as "woods" until the golf industry moved primarily to metal woods in the last years of the 20th century. In fact, the first metal woods made by TaylorMade, an early pioneer of that club type, were branded as "Pittsburgh Persimmons". Persimmon woods are still made, but in far lower numbers than in past decades. Over the last few decades persimmon wood has become popular among bow craftsmen, especially in the making of traditional longbows. Like some other plants of the genus *Diospyros*, older persimmon heartwood is black or dark brown in color, in stark contrast to the sapwood and younger heartwood, which is pale in color.

CULTIVATION

The oriental persimmon, *Diospyros kaki L.*, is an easy-to-grow tree which is adaptable to much of North Carolina. The tree has a compact spreading growth habit and low maintenance requirements. The ornamental beauty of its orange fruit and bright red foliage in the fall makes it an attractive plant in the home landscape. The tree is winter-hardy in eastern North Carolina, as well as the Lower Piedmont and Coastal Plain areas. In general, mature persimmon trees do not tolerate temperatures below 10 F.

Oriental persimmon trees are relatively free from serious insect and disease problems. However, there are several insect pests that should be mentioned that may be a problem in some areas. There is a persimmon trunk borer that will tunnel into the trunk of young trees near the soil line. If detected early this insect can be removed with a sharp knife before extensive damage is caused to the tree. Another insect pest is the persimmon Phylloxera which is a small insect that feeds on the persimmon leaves. Although the leaves may be deformed from the feeding insect, control measures are usually not recommended. If problems do arise, submit a sample of the foliage and insect to your local county Extension agent for identification and control measures. The primary problem the home-orchardist faces with persimmon culture is alternate year bearing. Alternate bearing is common in most varieties and is related to factors such as crop load (too heavy or too light), tree age or vigor, soil moisture, and pollination requirements. The varieties listed in this leaflet do not require cross-pollination to ensure a crop.

There are two major types of persimmon fruit based on fruit astringency. One group, the non-astringent types, may be eaten while the fruit is still mature and firm. Fruit of the other group must be soft before astringency is lost. Non-astringent varieties should be planted for home use although astringent varieties can be grown in North Carolina. Seeds are generally not produced in the non-astringent fruit as in the native persimmons, but seeds may be produced in some years in some varieties.

Site Selection

Persimmons generally bloom late enough in the spring (mid-April) to avoid spring frosts. The trees grow well on a wide range of soils, although they grow best on loamy, well-drained soils. As with other fruit trees, a soil pH of 6.0 to 6.5 is optimum for tree growth.

Persimmon trees will withstand drought, but fruit size and yield are reduced. Also, adequate moisture is required to produce sufficient shoot growth and formation of flower buds for next year's crop. During extended dry periods, irrigation should be provided to the trees on a weekly basis.

Full sun with some air movement is recommended for persimmon trees in inland areas, although they will tolerate some partial shade. Persimmons grown in cooler areas should have full sun with protection from cooling breezes. As an attractive ornamental the tree fits well in the landscape. It does not compete well with eucalyptus.

Soil

Persimmons can withstand a wide rage of conditions as long as the soil is not overly salty, but does best in deep, well drained loam. A pH range of 6.5 to 7.5 is preferred. The tree has a strong tap root which may mean digging a deeper hole than usual when planting (when on *D. kaki* stock).

Planting

Prior to planting, be sure to take a soil sample for pH and nutrient recommendations. If lime is required to raise the pH, apply it in the late fall or early spring and incorporation to a depth of 12 to 18 inches is required. Also prior to planting, make sure that varieties adapted for your area are selected. Rootstock selection is important too. Oriental persimmons are generally grafted onto seedlings of the American persimmon, Diospyros virginiana. The main advantage to using American seedling rootstock is that these seedlings tolerate excessive moisture and drought quite well and are adapted to the climate of North Carolina; however they are prone to suckering which needs to be pruned out annually. There is a lack of uniformity of tree vigor and size associated with *D. virginiana* rootstocks.

Care must be taken when transplanting a persimmon tree because of its fragile root system. In general, trees should be planted at the same depth they grew in the nursery. The root system must never be subjected to freezing or drying conditions. To ensure good root growth after planting, water the trees immediately after setting them out and on a weekly basis thereafter if they receive no rainfall.

Like all fruit trees, persimmons require full sun to assure good tree and fruit growth, as well as fruit bud development. Trees should be spaced 15 to 16 feet apart.

Irrigation

Persimmon trees will withstand short periods of drought, but the fruit will be larger and of higher quality with regular watering. Extreme drought will cause the leaves and fruit to drop prematurely. Any fruit left on the tree will probably sunburn. Some 36 to 48 inches of water are needed annually, applied gradually in spring and tapering off in the fall. Hot inland areas may require 2 or 3 applications weekly, while coastal areas may need watering only once every 6 weeks, depending on the soil. If a drip system is is used, the emitters should be moved away from the trunk as the tree matures.

Fertilisation

Most trees do well with a minimum of fertilising. Excess nitrogen can cause fruit drop. If mature leaves are not deep green and shoot growth is less than a foot per year, apply a balanced fertiliser such as a 10-10-10 at a rate of l pound per inch of trunk diameter at ground level. Spread the fertiliser evenly under the canopy in late winter or early spring.

Persimmon trees respond to nitrogen applications, especially in years of a heavy crop load. A rule of thumb is that 1 oz of actual nitrogen per year of tree age should be applied, although this should be adjusted according to the native fertility of the soil and the tree's general vigor. For example, a tree grown in a clay loam will require less applied nitrogen than a tree grown in a sandy soil. As a note of caution, excessive nitrogen can cause high amounts of premature fruit drop. Occasionally, other nutrients such as phosphorus, potassium or zinc may need to be supplied. A soil test will be necessary to determine exact recommendations. Contact your local County Extension agent for detailed instructions concerning soil testing.

Pruning

Prune persimmon trees to develop a strong framework of main branches while the tree is young. Otherwise the fruit, which is borne at the tips of the branches, may be too heavy and cause breakage. A regular program of removal of some new growth and heading others each year will improve structure and reduce alternate bearing. An open vase system is probably best. Even though the trees grow well on their own, persimmons can be pruned heavily as a hedge, as a screen, or to control size. They even make a nice espalier. Cut young trees back to 1/2 high (or about 3 feet) at the time of planting.

Persimmon trees are generally trained to a modified central leader system. Develop 3 to 5 well-spaced branches (approximately 8 to 12 inches apart) along the trunk. The first limb should be 3 to 4 ft from the ground. Then leave an area of approximately 2 ft on the leader without branches and allow 3 to 4 more branches and then cut off the leader above the highest branch.

Pruning mature trees should be kept at a minimum. Remove crowded or weak branches, diseased or broken limbs, or excessively vigorous growth such as watersprouts or root suckers. Pruning is required to renew fruiting wood since flowers are borne on current season's wood. New growth can be stimulated with moderate pruning cuts every 1 to 2 years. Pruning will also give a strong framework for the tree to bear a heavy crop load. This is important because persimmon wood is brittle and breaks easily with heavy crops.

Propagation

Stratification is recommended for all persimmon seeds. The common rootstock in California is *D. lotus*, although it is not compatible with some cultivars, including fuyu. Other rootstock such as *D. kaki* seedlings are temperamental and have long tap roots. *D. virginiana* is inconsistent and suckers badly. Whip and cleft grafts are the ones commonly used. The trunks of young trees should be protected from sunburn and rodent damage.

Pests and Diseases

Persimmons are relatively problem-free, although mealybug and scale in association with ants can sometimes cause problems. Ant control will usually take care of these pests. Other occasional pests include white flies, thrips which can cause skin blemishes and a mite that is blamed for the "brown lace collar" near the calyx. Waterlogging can also cause root rot. Vertebrate pests such as squirrels, deer, coyotes, rats, opossums and birds are fond of the fruit and gophers will attack the roots. Other problems include blossom and young fruit shedding, especially on young trees. This is not usually a serious problem, but if the drop is excessive, it may be useful to try girdling a few branches.

Over watering or over fertilisation may also be responsible. Large quantities of small fruit on an otherwise healthy tree can be remedied by removing all but one or two fruit per twig in May or June.

Harvest

Harvest astringent varieties when they are hard but fully colored. They will soften on the tree and improve in quality, but you will probably lose many fruit to the birds. Astringent persimmons will ripen off the tree if stored at room temperature. Nonastringent persimmons are ready to harvest when they are fully colored, but for best flavor, allow them to soften slightly after harvest. Both kinds of persimmons should be cut from the tree with hand-held pruning shears, leaving the calyx intact Unless the fruit is to be used for drying whole, the stems should be cut as close to the fruit as possible. Even though the fruit is relatively hard when harvested, it will bruise easily, so handle with care.

Mature, hard astringent persimmons can be stored in the refrigerator for at least a month. They can also be frozen for 6 to 8 months. Nonastringent persimmons can be stored for a short period at room temperature. They will soften if kept with other fruit in the refrigerator. Persimmons also make an excellent dried fruit. They can either be peeled and dried whole or cut into slices (peeled or unpeeled) and dried that way. When firm astringent persimmons are peeled and dried whole they lose all their astringency and develop a sweet, datelike consistency.

Fruit should not be picked until full color has been attained. The fruit bruises easily, so care should be taken when picking them. After harvest the fruit may be stored for several weeks in a refrigerator. Fruit can also be frozen and stored for longer periods. Prior to freezing, the fruit can be peeled, pureed, and put into tightly sealed plastic bags or containers. They can also be frozen whole, then thawed and processed as needed. The freezing process does remove some astringency, so the fruit do not need to be fully softened prior to freezing for the astringent varieties. Simply freeze the fruit until solid and remove fruit when desired.

The softening and ripening process can also be accelerated by placing firm, ripe (fully colored) persimmons in a loosely closed plastic bag with an apple or banana. Apples and bananas produce a gas, ethylene, which promotes fruit softening and ripening. Fruit can also be placed in a sealed bucket with a 2-inch x 2- inch piece of dry ice for 24 hrs and then opened and resealed with another piece of dry ice. The carbon dioxide from the dry ice will remove the astringency without softening the fruit.

Utilisation

Persimmon fruit are delicious "as is" fresh, or they can be used in a variety of baked goods such as puddings, breads, and cookies. The non-astringent types, such as 'Fuyu' and 'Jiro', may be consumed when the fruit have fully colored. The non-astringent fruit can be eaten while still firm. The flavor and texture are quite pleasant.

Farm Facilities

Persimmons can be successfully grown on a wide range of soils from light sandy loams to heavy alluvial clays. Preferred soils are light, well-drained sandy loams or loams with a minimum depth of 500 mm before clay or rock is encountered.

Where the topsoil is less than the minimum depth, mounding may be used to increase effective soil depth. Where possible, avoid acid soils naturally high in manganese because of the susceptibility of persimmons to green blotch disorder.

Slopes of up to 15% are suitable as machinery can be operated safely allowing a range of layout options while minimising soil erosion. Steeper slopes present a major erosion risk and make it difficult to operate machinery safely. These slopes also require expert assistance in block layout and design, which is very expensive to develop and maintain.

Slopes facing from north and north-east are preferred as these are warmer and more protected from damaging south-easterly and westerly winds. Windbreaks are recommended on all sites to minimise fruit abrasion, and assist orchard establishment.

The optimum climatic conditions for the production of high quality fruit are given below.

Climatic factor	*Explanation notes*
At least 1400 hours of sunshine during the growing season from October to April	This amount of sunshine produces fruit with the best colour and sugar levels
Mean daytime temperatures of between 15 and 22°C are necessary during the autumn fruit ripening period from March to May	Mean daytime temperatures less than about 14°C may result in only a partial loss of astringency in normally non-astringent varieties. Daytime temperatures above 25°C may affect fruit quality, and above 35°C may produce severe sunburn of fruit
Winter chilling of at least 100 chill units is required (preferably 350-550)	Although persimmon is a deciduous plant, and enters a rest period in winter, they appear to have a low chilling requirement for even bud break. They break bud in September, flower in October and are therefore less likely than other deciduous fruits to suffer flower and fruit damage from late spring frosts. Although low chill areas provide an opportunity for slightly earlier crops, high chill areas produce the best quality fruit
Minimal rainfall during the main fruit development and ripening periods	This minimises disease problems and other blemishes on the fruit

Protection from strong prevailing winds is essential	Fruit is very susceptible to wind damage and blemish. Windprotection is essential on all sites
Absence of frost at flowering time (mid October) is required	Plant damage is avoided

Irrigation is essential for the production of quality fruit. Water storage reserves to cover usage of 5 to 6 megalitres per hectare (ML/ha) are recommended. This may be reduced to 4 ML/ha in wetter areas but needs to be increased to 8 to 12 ML/ha in drier inland areas such as the Central Burnett and the Riverland and Sunraysia regions of southern Australia. As persimmons are sensitive to salt damage, irrigation water should not exceed 0.65 deciSiemens per metre (dS/m) and dissolved ions should be less than 400 milligrams per litre.

Machinery

This is a list of the essential equipment needed:

- permanent watering system such as under-tree mini-sprinklers or twin row drippers:
- a tractor large enough to operate spray equipment and narrow enough to fit through the orchard if growing trees on the trellised palmette system. A 4WD tractor is recommended for orchards on steep slopes;
- weedicide sprayer (knapsack, handgun or spray boom);
- power sprayer for pests and diseases (air blast sprayer preferred) - a smaller powered handgun sprayer can be used for small orchards or for the first few years in larger orchards but good spray coverage is difficult to achieve;
- splatter sprayer for applying fruit fly bait sprays;
- protective equipment for use when spraying;
- slasher or mower;
- trailer, carryall or utility for farm transport of fruit;
- shed for storage of equipment;
- chemical storage area;
- workshop space and tools.

Optional equipment includes:

- fertiliser spreader;
- mulch spreader;

— pneumatic/hydraulic secateurs.

When fruit production commences, the following equipment is also required:

— harvesting equipment including picking poles or ladders and picking buckets or bins;

— shed or undercover area with grading equipment (if packing own fruit);

— cool room preferably with forced air cooling facility (if packing own fruit)

— nets for protection against birds and fruit bats.

Small orchards may reduce capital costs by becoming part of a packing cooperative and by purchasing second-hand equipment.

Labour

Two people can comfortably handle up to one hectare of persimmon trees without the need for permanent labour. As the crop is generally grown in conjunction with other enterprises, this enables time for other necessary farming activities.

However, casual labour will generally be required to assist with pruning, thinning, harvesting and packing during the peak of the season. Pruning can take up to 100 hours per hectare over a two month period, fruit thinning up to 200 hours per hectare over a two week period, and harvesting up to 800 hours per hectare.

The number of casual workers required may vary from two to three for thinning, and up to seven to eight for harvesting and packing. Access to a central packing shed greatly reduces the need for on-farm labour for post-harvest treatment, grading and packing.

Peak harvest time is February to May. Heavy, manual labour is involved in almost every aspect of production. Particular care and training is needed to avoid injuries to workers and damage to fruit before and after harvest.

It is useful to record the cycle of tree growth, flowering and fruit production.

Major orchard operations include tree planting, fertilising, weed control, pest and disease monitoring, spraying, pruning, harvesting and packing. To be profitable, an integrated crop management approach is recommended. This requires willingness to either employ specialised consultants for pest monitoring or water management, or to learn these monitoring systems yourself.

Early maturing varieties of persimmon are very susceptible to fruit fly, and regular bait or cover sprays may be necessary from November to harvesting. Regular checking is needed for other pests to determine when control measures are necessary.

During peak harvesting, large quantities of fruit must be handled in a short period. This requires good organisational and labour management skills.

To keep the orchard at peak production with quality fruit, the orchard must be carefully managed through pruning and other cultural techniques. This requires special management skills.

Cultivars

There has been a great deal of confusion and misidentification among persimmon cultivars.

Astringent varieties

Eureka

Medium to large oblate fruit, puckered at the calyx. Skin bright orange-red. Good quality. Ripens late. Tree small, vigorous,drought and frost resistant, precocious and heavy-bearing. One of the most satisfactory cultivars for Florida and Texas.

Hachiya

Large, oblong-conical fruit Skin glossy, deep orange. Flesh dark yellow. Sweet and rich. Good for drying. Ripens midseason to late. Tree vigorous, upright-spreading. Prolific in California.

Honan red

Small, roundish oblate fruit with thin skin. Skin and flesh ripen to a distinct orange-red. Very sweet and rich. Excellent for fresh eating and drying. Ripens midseason to late. Tall, upright, moderately vigorous tree. Bears good crop.

Saijo

Small, elongated fruit. Skin dull-yellow when mature. Flavor sweet, excellent, ranked among the best by gourmets. Mature fruits are attractive when dried. Tree medium in height, bears consistently. Cold hardy to -10° F.

Tamopan

Large, somewhat four-sided fruit, broad-oblate and indented around the middle. Skin thick, orange-red. Flesh light orange, sweet and rich when fully ripe.

Tanenashi

Medium-sized round-conical fruits. Skin light yellow or orange, turning orange-red, thick. Flesh yellow, sweet. Ripens early. Tree vigorous, rounded, prolific. In California tends to bear in alternate years.

Triumph

Sold as Sharon Fruit after astringency has been chemically removed. Medium-sized, oblate fruits. Ripens in October.

Nonastringent varieties

Fuyu (Fuyugaki)

Medium-large oblate fruit, faintly four-sided. Skin deep orange. Flesh light orange, sweet and mild. Ripens late. Keeps well and is an excellent packer and shipper. Tree vigorous, spreading, productive. Most popular nonastringent cultivar in Japan.

Gosho/Giant Fuyu/O'Gosho

Large, roundish-oblate fruit. Skin reddish orange, attractive. When fully ripe has one of the deepest red colors of any persimmon. Flesh quality good, sweeter than Fuyu. Ripens in late October. Tree somewhat dwarf. Bears regularly but sets a light crop in some seasons and is prone to premature shedding of fruit.

Imoto

Similar to Jiro. Reddish brown skin. Occasional male flowers and seeds. Probably a bud mutation of Jiro. Ripens late October and early November

Izu

Medium-sized fruit. Skin burnt orange. Flesh soft, with a good amount of syrup, of fine texture. Flavor very good. Not reliably nonastringent. Ripens early, from the end of September to mid-October.

Jiro

Fruit large. Resembles Fuyu, but more truncated and squarish in cross-section. Skin orange-red. Flavor and quality excellent. Ripens late October and early November, ships well. Often sold as Fuyu. Tree slightly upright. Most popular nonastringent variety in California.

Maekawajiro

Medium-sized, rounded fruit, smoother and less indented than Jiro. Rich orange in color.

Sweet and of good quality. Ripens in mid-season. Tree slightly upright. Must be planted with a suitable pollinator to ensure good fruit yield. Bud mutation of Jiro.

Okugosho

Medium-sized, round fruit. Skin orange to deep red. Flesh sweet, of good texture, flavor good. Not reliably nonastringent. Ripens in early November. Tree medium-sized, vigorous, spreading. Differentiates male flowers, making it a suitable pollinator.

Suruga

Large fruit. Skin orange-red. Flesh dense, very sweet, excellent quality. Difficult to soften on tree (fruit becomes spongy rather than soft). Ripens in November, keeps well Tree almost free from alternate bearing. Recommended for warmer climates.

Pollination Variant Varieties (astringent when seedless)

Chocolate

Small to medium-sized, oblong-conical fruit. Skin reddish orange. Flesh brown-streaked when pollinated, must be soft-ripe before eating. Ripens late October to early November. Tree large, vigorous, producing many male blossoms. Recommended as a pollinator for pollination variant cultivars such as Hyakuma and Zenji Maru.

Gailey

Fruit small, roundish to conical with a rounded apex. Skin dull red, pebbled. Flesh dark, firm, juicy, of fair flavor. Tree small to medium. Bears many male flowers regularly and is an excellent cultivar to plant for cross-pollination. Has attractive autumn foliage and ornamental value.

Hyakume

Fruit large, roundish oblong to roundish oblate. Skin buff-yellow to light orange, marked with rings and veins near the apex. Flesh dark cinnamon when seeded, juicy, of firm texture, nonmelting. Flavor spicy, very good. Nonastringent even while the fruit is still hard. Ripens in midseason, stores and ships well.

Maru

Small to medium-sized fruit, rounded at the apex. Skin brilliant orange-red, attractive. Flesh dark cinnamon, juicy, sweet and rich, quality excellent. Stores and ships especially well. Tree vigorous and productive. Generally considered a group name.

Nishimura wase

Fruit medium, round conical to oblate. Orange color. Mediocre flavor. Ripens in September. Bears male flowers.

Japanese Persimmon Varieties

Among (Yemon)

Large, two and one-half to three inches in diameter, round, flattened, deeply ribbed; color dark orange red or yellowish red. Flesh red, very sweet, and edible while quite firm, but quality improves as it becomes soft. Season, September to November.

Costata

Medium, conical, pointed, somewhat four-sided; color yellow. Flesh light yellow; dark flesh unusual; seeds rare; astringent until it softens, when it is very good.

Hachiya

Large to very large, two and one-half by three inches in diameter, oblong, with blunt apex, somewhat ribbed; color bright crimson, with lines and blotches at apex. Flesh orange red, sweet and good when soft, but astringent when solid. Needs house ripening, and can be kept late. One of the varieties used for drying.

Hiyakume

Large, three inches in diameter, usually flattened but often elongated; color yellow, with netted lines at the apex. Flesh orange red; only used when soft; a remarkable keeper. Tree of dwarf growth.

Ioyama Gaki

Medium to large, round slightly conical; color yellowish orange, with dark netting at apex. Flesh dark brown, very sweet, and good when eaten before softening.

Kurokuma

Very large, three to three and one-half inches in diameter, roundish, flattened. Flesh red, sweet, and good when soft; keeps late, and requires house ripening.

Minokari (Large Stork Egg)

Large to very large, oblong, pointed; color vermilion red. Flesh red, very good in quality, but must be ripened indoors; valuable for drying.

Mio-Tan (Mazelli)

Medium in size, roundish, somewhat oblong, slightly ribbed; color orange red. Flesh brownish red, but variable.

Keeps late, but the brownish specimens are edible and good before softening. Extensively grown in South France.

Okame

Large, roundish oblate, with netted lines at apex; color orange yellow, changing to carmine, with bloom and waxy expression. Flesh light-colored, with light brown center around the seeds. Loses its astringency when well colored.

Taber No. 23

Medium, oblate, flat or depressed at apex; color dark red with peculiar dotted surface. Flesh dark brown, sweet, and not astringent; quality good.

Taber No. 129

Medium, roundish, flattened at base, with point at apex; color dark yellowish red, with quite rough surface. Flesh light brown, crisp, sweet; free from astringency. Keeps well and ships well.

Tane (Nashi)

Large to very large, diameter three and one-quarter by three and three-eighths; roundish conical, pointed at apex; color light yellow, changing to bright red. Flesh yellow and seedless; quality fine.

Triumph

Medium, tomato-shaped; color yellow; has few seeds; quality very good. Said to be very productive.

Tsuru-No-Ko (Stork Egg)

Large, two and one-half by three and one-half inches, oblong, pointed; color bright red, sometimes with black at apex. Flesh red, very good. Must be ripened in the house.

Yedoichi (Maru-Gata)

Medium, round, often slightly oblong; flattened at base and narrowed at apex; color dark red, often with black mottling near apex. Flesh mahogany brown, with darker spots. Edible while solid. Fruit in large clusters.

Zengi

Small, one and three-quarters by two inches, round or oblate; color light, with reddish shades. Flesh dark brown, with darker spots; very sweet and good; edible while solid in September and has a long season through October.

If the fruit is to be shipped to distant markets, it should be gathered when fully grown but before it has begun to soften. Clippers or picking-shears should be used, and the fruit must be handled carefully, since it is easily bruised. Even when intended for home use it is preferable to gather it. before it has begun to soften, and then ripen it in a dry warm room. Fruit treated in this manner is fully as good as that ripened on the tree.

MARKETING

In a modern horticultural business, an understanding of marketing and a commitment to quality throughout the entire production and marketing system are essential. This involves regular communication with people in the market chain, and a willingness to work with other growers in cooperative marketing ventures.

To be successful, the orchard must be run as a business. This is a complex operation requiring many skills including business planning, bookkeeping, and maintaining farm records. As production and marketing technology changes, it will also help if you are prepared to experiment with and implement new ideas. There are several options for marketing your fruit. They include:

- individual growers, grower groups, or cooperatives selling direct to central produce markets in Australia's major cities;
- direct selling to major city chain stores and fruit barns;
- selling locally.

Traditionally growers send their fruit to their preferred agent or agents for sale, but cooperatives and marketing groups have started so that growers can play a more active role in distributing and marketing their fruit.

There are two types of persimmon grown in Australia -astringent and sweet (non-astringent) persimmon. The commercial industry is focused almost entirely on the sweet (non-astringent) types, whereas it is very common for astringent persimmon to be grown in backyards. Unfortunately many consumers confuse the two types and do not understand that commercially grown varieties are sweet (non-astringent).

A future challenge for the industry is to introduce sweet (non-astringent) persimmon to more Australian consumers. Asian consumers are much more familiar with both sweet and astringent persimmons.

Australia grows about 350 000 trees of non-astringent cultivars with an estimated fruit production of 450 000 trays. This is equivalent to an annual production of 2 500 tonnes. In comparison with world production of 600 000 tonnes, Australian production is relatively small (less than 0.5% of world production). Average orchard size is about 440 trees.

Tree yields range from 5 to 35 tonnes per hectare due to improved orchard management. Average yield reported in one survey in 1999 was nearly 16 tonnes per hectare. Only 40% of the plantings are freestanding and the rest are trellised palmette.

The main sweet persimmon production areas are southern Queensland, northern New South Wales, Coffs Harbour, Sydney, northern Victoria (Cobram, Shepparton, Sunraysia), north eastern South Australia (Riverland), and southwest Western Australia.

Queensland grows about 200 hectares of sweet persimmon mainly in the following areas: Sunshine Coast , South Burnett, Toowoomba Range , Lockyer Valley , Childers/ Bundaberg and the Granite Belt. Small isolated plantings exist in other coastal areas and on the Atherton Tablelands. It is estimated that the Queensland industry makes up about 40% of the Australian industry.

The modern industry is comprised almost entirely of sweet Japanese persimmon varieties. Fuyu is the most commonly planted variety in Australia (about 70%), followed by Jiro (15%), Izu (10%), and Suruga (5%). These newer non-astringent varieties yield very attractive sweet, orange-red fruit, which can be eaten in a hard mature condition. These characteristics, together with Asian consumer familiarity, suggest that sweet persimmon has significant consumer potential, both domestically and internationally.

Persimmon trees are grown from grafted nursery stock where the cultivar is grafted onto a selected rootstock. Trees have high yielding potential and are normally grown on trellises in intensive orchard blocks. Over-tree netting is important in protecting fruit from bird and fruit bat damage.

Australia's production period starts in the warmer semi-tropical regions of Queensland in late February and finishes by late June in the cool temperate regions of southern Australia. The majority of fruit is sold through the major metropolitan wholesale markets in Brisbane, Sydney and Melbourne, but a significant quantity is exported to southeast Asia. Prices in all major Australian cities are extremely variable reflecting the variability in quality.

Good quality larger fruit may receive $10-18 per tray during the season. In contrast, poor quality fruit may receive as low as $3 per tray. New Zealand fruit, our major competitor in the Australian and export markets, is available from late May to August.

It is estimated that less than 10% of people have eaten non-astringent persimmon, so there is considerable potential for this fruit. Asian consumers like the sweetness and low acidity of persimmon.

Exports of persimmon have steadily increased from virtually nil in 1990 to over 100 000 trays in 2001.

The Australian Persimmon Export Company (APEC) is the largest Australian company exporting about 60 000 trays per year. Major export markets are Singapore and Malaysia, with smaller markets in Thailand and Hong Kong. Quality and consistency of export fruit have been a problem limiting expansion. Prices for high-quality fruit in these markets is high varying from $15 to $20 per tray. In comparison, the exports of New Zealand fruit has risen to 600 000 trays over the same period.

The major fruit quality problems have been rapid softening of fruit during transit and storage, immature fruit, and uneven and poor colouring. Australian fruit is sweeter (more than 14° Brix) than New Zealand fruit.

Fruit Quality

Domestic fruit quality problems vary within regions and States. Rapid softening of fruit in transit and storage continues to be a major problem from most regions, except South Australia.

In the past, the percentage of marketable fruit has been reduced by as much as 60% due to blemish marks (wind rub, petal adherence marks, sunburn, spots and speckles), calyx separation and apex cracking. Training trees onto trellis systems can significantly reduce the amount of blemish.

Fruit grown in coastal regions of Queensland and northern NSW do not develop a deep orange skin colour because of the lack of cooler night temperatures. By contrast, in South Australia , fruit colour is excellent but fruit size may be reduced up to two size grades because of stress caused by saline water and soils.

Persimmons are an attractive crop for new growers since trees come into production relatively quickly and the crop is adaptable to a wide range of climatic environments. It also has a high potential yield per unit area, but current yields on many orchards are low due to poor management practices.

Growers are attracted to the fact hat the fruit has characteristics that give it significant growth potential for both domestic and export marketing. However, be aware of the following constraints:

- The sweet persimmon is still largely unknown in many markets and significant promotion and marketing development will have to be done to ensure that it is not affected by the stigma of past adverse consumer experiences with older astringent persimmons. This development will be necessary to ensure that growers continue to receive reasonable prices in view of the significant increases in market volume expected in the next few years.

- Although the crop has high potential yield per unit area, a high proportion of the crop will be of inferior quality unless the trees are trellised, and high levels of crop management applied. This requires skill and is expensive and time consuming. The need for pruning, fruit thinning and harvesting of a large crop over a short period of time means the crop is also very labour intensive. The impact of this labour requirement can be gauged from the fact that most commercial orchards consist of between 500 and 3000 trees.
- Persimmons are very susceptible to bird and fruit bat damage and netting is considered essential in most areas to achieve good production of quality fruit.
- Because of the need for trellises and netting, the crop requires a high initial capital investment.
- The persimmon industry is relatively new and small compared to other horticultural crops, so there is only a limited range of chemicals registered for pest and disease control. This may make it difficult to control these problems in the future.

For these reasons, it is recommended that new growers carefully analyse the market and prepare a thorough business plan before proceeding. Thoroughly research all of the above issues with other growers, marketers, wholesalers, industry experts and consultants as you develop your business plan.

Persimmons are a relatively new crop and the lack of information on the economics of growing them put them in a higher risk category. Having said that, there are many good growers in the industry. Persimmon is not normally grown as the sole source of income. We recommend that new growers start with no more than one hectare of trees, learn how to manage these properly, and then increase their planting from there. For these reasons, the economic figures that follow are based on an orchard of one hectare of trees.

Yields can vary widely, but three year old trees on a close planted palmette trellis (approximately 1000 trees/ha) should yield about 7 kg of fruit and mature trees from six years about 30 kg. This equates to a yield of about 30 tonnes per hectare for a well-managed, mature orchard on a palmette trellis.

For average quality fruit, prices range from about $18 per 4 kg tray for early season Izu fruit, to an average of about $10 per tray in April in the peak of the season. Please note that as more farms come into production early season prices are declining. Production costs are estimated to be in the range $8.50 for larger, well managed orchards up to $11 per tray for smaller, less productive orchards.

No income is received until about the third year and annual expenses exceed annual income until the fourth to fifth year. Accumulated income should exceed accumulated expenses by the sixth or seventh year. This is a short payback period compared with other tree crops.

At maturity, one well-managed hectare should conservatively return about $12 000 after production costs are covered. This is based on an average yield of 4000 trays per hectare, average prices of $12 per tray and costs of $9.00 per tray. This covers all growing and marketing costs such as fertilisers, pest and disease control, casual labour, fuel, packing and wholesale merchant's fees. No allowance has been made in these calculations for house and land purchase, loan repayments, living expenses, capital items or fixed costs such as rates, taxes and depreciation of machinery and equipment.

Excluding the cost of house and land, it has been estimated that about $120 000 is needed to establish a one hectare sweet persimmon orchard. This covers the cost of basic machinery, an irrigation system, a shed, land preparation, trellises and tree establishment. Establishing one hectare of trees, with just irrigation, netting and V-trellising systems is in the vicinity of $32 000 to $47 000. These figures assume a cheap removable netting system is erected each year when the crop is susceptible to bird and fruit bat damage, permanent netting structures will add considerably to these costs.Further funds would be required in the second year, when trees begin bearing, to purchase a forced-air cold room, a fruit grader, and miscellaneous harvesting equipment.

REFERENCES

Carley Petersen and Annabelle Martin. "General Crop Information: Persimmon". University of Hawaii, Extension Entomology & UH-CTAHR Integrated Pest Management Program. http://www.extento.hawaii.edu/kbase/crop/crops/i_persim.htm. Retrieved on 2007-01-15.

Mish, Frederic C., Editor in Chief *Webster's Ninth New Collegiate Dictionary* Springfield, Massachuetts, U.S.A.:1984—Merriam-Webster Page 877

"Persimmon Fruit Facts". California Rare Fruit Growers, Inc.,. http://www.crfg.org/pubs/ff/persimmon.html. Retrieved on 2007-01-15.

8

Plum

A plum or gage is a stone fruit tree in the genus Prunus, subgenus Prunus. The subgenus is distinguished from other subgenera (peaches, cherries, bird cherries, etc) in the shoots having a terminal bud and the side buds solitary (not clustered), the flowers being grouped 1-5 together on short stems, and the fruit having a groove running down one side, and a smooth stone.

Plums are popular for cooking, jam making and bottling or canning, but the sweeter varieties are among our most delicious dessert fruits. Damsons ripen a little later than most plums. The fruits are small, oval and richly flavored, but not really sweet enough for the general taste for eating raw. They are, however, excellent for cooking, preserves and bottling. Bullaces are small round fruits, which ripen even later and are useful on that account to lengthen the season. Bullaces can be eaten raw but are excellent for cooking. Gages are simply a class of plum with a characteristic, and particularly delicious, flavor. Gages, bullaces and damsons are all grown in the same way as plums.

Plums will grow in most parts of the country but as they flower early they are very vulnerable to spring frosts. The choicer kinds deserve the protection of a wall where protection from frost (and birds) can more easily be given. They do best in areas where the annual rainfall is between 50 and 90cm (20 and 35in). Damsons will succeed in areas having higher rainfall, and less sunshine, than plums will tolerate.

Plums need a well-drained soil and one containing plenty of humus to hold moisture during the growing season. A very acid soil should be limed, but an alkaline soil should not be planted with plums. Plums (and other stone fruits) do need calcium but they will not prosper in an alkaline soil. Plum trees planted in thin soils overlaying chalk often suffer seriously from lime-induced iron deficiency.

No really satisfactory dwarfing rootstock has yet been found for plums. The two least vigorous are common plum and St Julien 'A'; the former, however, is only compatible with certain varieties. Trees grown on these rootstocks are sometimes described as 'semi-dwarf' but, even so, a standard or half-standard would be too large for the average garden, and even a bush-type tree requires a spacing of 4-5m (12-15ft) (on Brompton or Myrobalan 'B' rootstock, 6-7m [18-20ft]).

As plums do not produce fruiting spurs as apples and pears do, they are not so amenable to training, and are seldom satisfactory as cordons or espaliers. They may, however, be grown as fans, for wall-training or with the support of posts and horizontal wires, but root-pruning will probably be necessary every five years or so to restrain growth and maintain fruiting. A fan tree on St Julien 'A' rootstock should be allotted at least 5m (15ft) of wall space.

Plums may also be grown as semidwarf pyramids on St Julien 'A' rootstock and this is a form, which is best for the small garden. Such a tree requires a spacing of 3.3m (10ft) and, as it will never be allowed to grow much over 3m (9ft) in height, it is possible to arrange some kind of cage or netting over the top of the tree to keep off birds, which will otherwise damage the fruit. An additional advantage is that the branches of a pyramid seldom break and there is thus less likelihood of infection by disease.

For training as a pyramid a maiden should be planted in the usual way and the following March it should be headed back to 1.6m (5ft). Any laterals above 45cm (18in) from soil level should be shortened by half and any arising lower down the stem should be cut off entirely. Towards the end of July or early in August, when new growth has finished, cut back branch leaders to 20cm (8in), making the cut to a bud pointing downwards or outwards. Cut laterals back to 16cm (6in). Repeat this procedure annually. Leave the central leader untouched in summer but in April of the second year cut it back to one-third of its length. Repeat this annually, cutting the new growth back by two-thirds until a height of 3m (9ft) is attained. After that shorten the new growth on the central leader to 2.5cm (1in) or less each May.

Plant plums in the usual way between November and March, the sooner the better, always provided the soil is friable. Stake securely and put down mulch to preserve soil moisture.

An established plum needs plenty of nitrogen but, until good crops are being carried, on most soils it will be sufficient to give a light mulch of rotted farmyard manure or garden compost in spring, and prick this lightly into the surface the subsequent autumn. When good crops are being borne, the yearly mulch may be supplemented with 28g (1oz) per dressing of Nitro-chalk and 14g (0.5oz) per sq. m sq. yd) of sulphate of potash, given in February. Every third year, add 28g (1oz) per sq. m sq. ft) of superphosphate. Where no manure or garden compost is available, peat may be used as mulch and the dose of Nitro-chalk doubled.

The wood of plum trees naturally tends to be brittle and branches often break in late summer gales when the crop is heavy. Thinning of the fruit will help to prevent this form of breakage, and it is also advisable to arrange some kind of support for extra-heavily laden branches on bush-type trees. Wooden props may be fixed beneath branches or a tall, strong central pole can be erected and branches supported from this by ropes, maypole fashion.Dessert plums should be left on the tree until quite ripe and then picked by taking hold of the stalk so that the place. They will keep for a couple of weeks or so.

There are thousands of varieties of plum, all differing in size, shape, colour and flavour. These members of the rose family originate from three main types - European, Japanese and Western Asian. The skins can vary from blue-black to purple, red, green and yellow. They have a long season, and one variety or another is available almost all year round. All plums have smooth skins with a bloom and juicy flesh with plenty of acidity.Dessert plums can be eaten on their own; they are usually larger than cooking plums (up to 10cm/4in long) and are sweet and very juicy. Cooking plums are drier, with tart flesh that is ideal for pies, flans and cakes.

Plums (*Prunus domestica*) date back in writing to 479 B.C. in the writings and songs of Confucius which include a listing of popular foods of Chinese culture. In 65 B.C., Pompey the Great introduced the plum to the orchards of Rome, and Alexander the Great eventually brought them to the Mediterranean regions. Early American colonists found wild plums growing along the east coast, but today the common European plum has replaced the native wild plum in popularity and as a commercial crop. The plum tree plays a significant role in Chinese mythology and is associated with great age and wisdom. Blossoms of the plum tree are carved on jade to signify resurrection. Plums are now the second most cultivated fruit in the world, second only to apples.

Wild plums originated in Asia at least 2,000 years ago. They were first cultivated by the Assyrians, then adopted by the Romans, who hybridised them with great enthusiasm; the historian Pliny wrote of the huge numbers of plum cross-breeds available. The Crusaders brought plums to Europe, where they became highly prized. Nowadays they are grown in almost all temperate countries.

Plums are known as prunes in France, and earlier times, it is believed that both plum and prune meant the fresh fruit. Of course, in modern times, the prune is the dried version of the plum. Most prunes are made from the La Petite d'Agen variety of plum brought from France in 1856 by French horticulturist Louis Pellier. A new movement by the prune industry is to market prunes as plum raisins or dried plums in hopes the new term will appeal more to younger people.

There are over 2,000 varieties of plum, ripening at different times throughout summer and autumn, although only a dozen or so are available in the shops. Japanese varieties are large, round and juicy; they can be purplish-red with orange flesh, or orangey-yellow with yellow flesh. On the whole, dark-coloured plums have bitter skins, while the red and yellow varieties tend to be sweeter. Most dessert plums can be cooked as well as eaten raw.

CULTIVATION

Plum fruit is sweet and juicy and it can be eaten fresh or used in jam-making or other recipes. Plum juice can be fermented into plum wine; when distilled, this produces a brandy known in Eastern Europe as Slivovitz, Rakia, Tzuica or Palinka. Dried plums are known as prunes. Prunes are also sweet and juicy and contain several antioxidants.

Figure 1. Plum fruits

Prune marketers in the United States have, in recent years, begun marketing their product as "dried plums". This is due to "prune" having negative connotations connected with elderly people suffering from constipation.

Various flavours of dried plum are available at Chinese grocers and specialty stores worldwide. They tend to be much drier than the standard prune. Cream, Ginsing, Spicy, and Salty are among the common variety flavours. Licorice is generally used to intensify the flavour of these plums and is used to make salty plum drinks and toppings for Shaved Ice or *baobing*.

Pickled plums are another type of preserve available in Asia and international specialty stores. The Japanese variety, called *umeboshi*, is often used for rice balls, called "Onigiri" or "Omusubi". The *ume*, from which umeboshi are made, is however more closely related to the apricot than to the plum.

Prune kernel oil is made from the fleshy inner part of the pit of the plum.

Plums come in a wide variety of colours and sizes. Some are much firmer-fleshed than others and some have yellow, white, green or red flesh, with equally varying skin colour.

Plums and prunes are known for their laxative effect. This effect has been attributed to various compounds present in the fruits, such as dietary fiber, sorbitol, and isatin. Prunes and prune juice are often used to help regulate the functioning of the digestive system.

Plum cultivars in use today include:

— Damson, or Damask Plum

— Greengage (Firm, green flesh and skin even when ripe.)

— Mirabelle (Dark yellow, predominantly grown in northeast France.)

— Satsuma plum (Firm red flesh with a red skin.)

— Yellowgage, or Golden plum (Similar to Greengage, but yellow.)

When it flowers in the early spring, a plum tree will be covered in blossom, and in a good year approximately 50% of the flowers will be pollinated and become plums. Flowering starts after 80 growing degree days.

If the weather is too dry the plums will not develop past a certain stage, but will fall from the tree while still tiny green buds, and if it is unseasonably wet or if the plums are not harvested as soon as they are ripe, the fruit may develop a fungal condition called brown rot. Brown rot is not toxic, and very small affected areas can be cut out of the fruit, but unless the rot is caught immediately the fruit will no longer be edible. Plum is used as a food plant by the larvae of some Lepidoptera including November Moth, Willow Beauty and Short-cloaked Moth.

The Serbian plum is the third most produced in the world and the alcoholic drink slivovitz (Plum brandy) is the national drink of Serbia.

A large number of plums are also grown in Hungary where they are called szilva and are used to make lekvar (a plum paste jam), *palinka* (a slivovitz-type liquor), plum dumplings, and other foods. The region of Szabolcs-Szatmár, in the northeastern part of the country near the borders with Ukraine and Romania, is a major producer of plums.

The mei blossom (*Prunus mume*), along with the peony, are considered traditional floral emblems of China. On June 21, 1964, the Executive Yuan of the Republic of China officially designated the mei blossom to be its national flower, with the triple grouping of stamens representing the Three Principles of the People and the five petals symbolising the five branches of the ROC government. The designation repeats a previous statement by the ROC government in 1929.

The *mei* blossom is also the floral symbol of the ancient Chinese city Nanjing, which served as the former capital of the Republic of China.

SPECIES

The subgenus is divided into three sections:

- Sect. Prunus (Old World plums). Leaves in bud rolled inwards; flowers 1-3 together; fruit smooth, often wax-bloomed.
  - *P. cerasifera* (cherry plum)
  - *P. cocomilia*
  - *P. consociiflora*"
  - *P. domestica* (species of most "plums" and "prunes" sold as such)
  - *P. simonii*
  - *P. spinosa* (blackthorn or sloe)
- Sect. Prunocerasus (New World plums). Leaves in bud folded inwards; flowers 3-5 together; fruit smooth, often wax-bloomed.
  - *P. alleghaniensis*
  - *P. americana*
  - *P. angustifolia*
  - *P. hortulana*
  - *P. maritima* (beach plum)
  - *P. mexicana*
  - *P. nigra*
  - *P. orthosepala*
  - *P. subcordata* (Klamath, Oregon, or Sierra plum) Purple; fruit velvety. Treated as a distinct subgenus by some authors.
  - *P. armeniaca* (apricot)
  - *P. brigantina*
  - *P. mume* (ume)
  - *P. sibirica*

GROWING PRAIRIE PLUMS

Plum cultivation on the prairies has been occurring with varying degrees of success for over 80 years. There are wild plums (*Prunus nigra*) in Canada and these have used in breeding programs to increase winter hardiness in the Japanese plum (*P.salicina*). Much of the breeding work has been done at the University of Saskatchewan. There are a few guidelines that must be followed in order to have good success in growing plums in our hostile climate. Organic production is relatively easy for plums on the prairies and would be an advantage in their marketing.

Site Selection

Plum trees do well in a clay loam or sandy loam soil with at least one meter of soil below. A wet or hard grey soil is undesirable as the roots will suffer and die back can occur.

North or east facing slopes are ideal location for early blooming trees because delayed flowering time will aid in avoiding early frosts which can kill flowers. This type of orientation can also help prevent sunscald.

Firstly, plan the position of the plum tree according to its eventual size at maturity. Allow enough room so that it is not crowded.

Plums have three main needs, warmth especially important at pollination time, light and moisture. Plums flower earlier in the season than most other fruit trees, so they should not be planted in a frost-pocket. If your garden is in a cool area, avoid early flowering varieties - good varieties to choose are Czar, Victoria or Marjorie's Seedling. Plum trees prefer to be positioned in full sun, although some shade in the morning or afternoon will not affect them much.

As far as moisture is concerned, do not plant in a water-logged area, but make sure the soil is unlikely to dry out. Don't plant them near other trees which will simply deprive them of the moisture they need.

Micro Climate

Providing a shelterbelt on all sides of the orchard is essential in trying to modify your growing environment by reducing the flow of harsh winds. Shelterbelts also trap snow cover and can provide greater protection to the roots, as well as providing added moisture in the spring.

Planting and Establishing

Work the soil the year before planting trees. Plant a green crop and till into soil which helps make sure the area is weed free. Plant trees at 2-4 meters apart in rows depending on the system and which root stock used. Young fruit trees will need irrigation for the first 3 years until they have established a proper root system.

Technically, you can plant your plum tree any time from late Autumn to early Spring. The best time to plant is in October when the soil is moist but still retains some of the summer warmth.

Soil preparation is best done a month or more before planting so that the soil has time to settle. Dig a 60cm (2ft) deep by 1.2m (4ft) square hole, incorporating as much organic material as possible - aim to get the soil crumbly.

If the tree is being planted in soil which has previously been fertilised for other crops, do not add more fertiliser - too fertile a soil will result in too much tree growth at the expense of too little fruit growth. If the plum tree is being planted in a lawn, prepare as above, working in three handfuls of bonemeal or other long-lasting fertiliser.

Planting is simple - in basic terms, dig a hole large enough to easily take the roots, place the tree in the hole and cover the roots with soil up to the surrounding ground level. Simple enough, but bear in mind a few points.

Don't add any fertiliser to the soil at this time - it may burn the roots and it will only encourage tree growth at the expense of fruit growth.

The tree should be planted to the same depth as it was in the pot (or the soil mark on the trunk in the case of bare-rooted trees). If in doubt, make sure that the joining point between the rootstock and scion is at least 5cms (2in) above ground level. Having planted the tree, firm down the soil using your boots to ensure the soil is in good contact with the roots - water well if the conditions are dry.

Plum trees should be supported with a stake for the first couple of years of their life. The stake should be 15cm (6in) or so from the main stem. Tie the trunk to the stake at 30cm (1ft) intervals using plastic ties available from garden centres - do not use wire or anything which could cut into the tree trunk. The ties will need to be checked for the first couple of years to ensure that growth of the tree trunk has not caused them to become too tight.

Care of Plum Trees

With newly planted trees, check in their first winter that they are not lifted by frost, and provide them with water during dry spells. Plum trees grown against a wall or fence will need more regular watering throughout their life.

With more mature trees, very little care other than pruning and thinning is required. Your plum tree will definitely thank you for a layer of well-rotted compost spread around, but not touching, the trunk. Early spring is the best time for this, to help preserve water and keep weeds at bay.

An Autumn application of a long lasting fertiliser such as bonemeal will keep them well fed and ready for growth next year.

Because plum trees have lots of roots near the surface, they will often produce lots of of suckers (mini-trees) for 3 metres (9 foot) around the trunk. Regular cutting of these suckers will keep them under control. A more permanent way is to gently dig away the soil to the point where the sucker joins the main root system. Simply pull the sucker away from the root. This will slow down the production of suckers considerably.

Thinning Plums

Plum trees can be a bit erratic in producing fruit - one year they produce too little and the next year they produce too much. In years when they produce too much, it's best to remove some fruit before it matures to ensure a good crop next year and also better quality fruits this year. The other danger of producing too much fruit is that the branches may not be able to support their weight and they may break.

Thin the fruits at two stages in the year. First in late mid-May when the fruits are beginning to form, the again in July when the plums are about half their full size. First, remove any damaged or diseased fruit. Then, aim to thin the fruit so that there is enough room for the plums to develop without touching each other - about 7cm (3in) between each plum.

Supporting Branches

If the weight on some of the branches looks like it may cause them to break, it will be necessary to support the branch with a stake from below. Where this is done, pad the branch with some sort of material at the point it is supported by the stake. This will prevent the branch rubbing on the stake and maybe letting disease in.

Pruning Plum Trees

There are two considerations when pruning plum trees. The first is to establish a good shape. The second is to avoid any infection with silver leaf disease. The key to avoiding this infection is to prune plum trees at the correct time of year.

To avoid silver leaf disease, prune plum trees in June when they are growing strongly, do not prune in the winter. Try to make the pruning cuts as cleanly as possible, and avoid crushing the wood. If any larger sections of branch are to be pruned, seal the cut with a protective sealer available from garden centres. Burn any diseased or dead wood.

Pruning of plum trees depends on the form wanted - pyramid, bush, standard or half standard. Pyramid trees are pruned in a different way from the other forms.

Pruning in northern areas should be done in late winter or early spring in order to avoid winter injury. Summer pruning makes the tree vulnerable to diseases. One year trees should be headed back in the spring to encourage production of a strong and healthy main branch which will act as the central leader. The central leader system should be maintained each year by removing competitive leaders and excessive side branching. Removal of dead/disease wood and crossing branches is very important in maintaining a healthy and productive tree. Water sprouts should also be removed at base of tree because they are highly susceptible to fire blight invasion. If training to a trellis system, follow the trellis pruning system.

Choosing Your Plum Tree

Give careful consideration to what type of plum tree is best for you. Buying a plum tree which might grow to 9m (30ft) tall producing masses of cooking plums is not a good idea if you have a small garden and want eating plums!

The three key factors in choosing a plum tree are size, taste and pollination. GardenAction explains all three.

Size: Fan trained plum trees grow to an eventual height of around 2 m (7 ft). Although the shortest of all the forms, they will grow to a width of around 3 m and are really only a practical proposition when grown against a wall. If you have that sort of wall space available, this is the form for you. Pyramid trained plum trees are free-standing and will grow to an eventual height of 2.5 m (8 ft). These are a good form for a smallish garden - their spread will be about 1.6 m (5 ft). This form of tree has branches radiating out in random directions from a straight central trunk

Bush trained plum trees are free-standing and are taller than the pyramid form - around 4m when mature. They do have one advantage over the pyramid form however, the branches and foliage is held around 1 m (3 ft) high from the ground - this makes mowing around the tree much easier. The downside is that some of the fruit will be too high to pick without the use of a ladder.

Half Standard trained plum trees are again free- standing and will grow to a height of around 6 m (20 ft). This is a largish tree and only suitable for medium to large gardens. It will produce a mass of plums each year, most of them needing a ladder to harvest.

Standard plum trees reach a height of around 8 m (28 ft) at maturity - on your own head be it if you plant one of these in anything but a large garden. They will produce enough fruit to feed your entire neighbourhood and you will definitely need a decent sized ladder to pick most of the fruit.

Taste: The best way to make a decision on taste, is to understand a little about the variety of plums available.

Bullace—strictly a cooking plum. The trees are smaller than normal and very hardy, they are also ornamental. The fruit has a very sharp flavour, excellent for jams and preserving.

Damson -- another cooking plum, but sweeter than the bullace. The fruits are have a sharpish taste and are ideal for pies, tarts and jams.

Plum — some are cooking and some are eating. These make up some of the best plums for the UK climate.

Gage—eating (desert) plums. These are some of the the sweetest form of plum and they have a distinct 'plum' fragrance.

Pollination: Some plum tress are self-fertile, but many require a compatible plum tree nearby (plum trees are not so common as apple trees) for pollination to occur. Plum trees have a short and very distinct pollination period (almost exactly ten days) so if you choose a tree which is not self-fertile, be sure to also choose a compatible tree. This is especially important if you choose a 'gage'.

The majorities of prairie grown plums are self-incompatible or have low fruit production if not cross pollinated. This is overcome by having more than one plum tree cultivar in a planting area. Another way is to graft a suitable pollinator cultivar onto the existing tree. The hybrid plums are especially poor pollinators and should always be planted with a native cultivar.

Diseases

Black knot: Caused by the fungus *Dibotryon morbosum*. It will appear on current season's growth as a 1 cm brown corky scab. It usually appears in late summer and will continue to grow into a black warty mass not unlike dog feces.

Control: Prune all infected branches and burying or burning the infected wood. Make sure to prune back at least 10 cm below knot to stop the spread of disease within the branch.

Fire Blight: All members of the sub-family Maloideae are susceptible to fire blight, although plums are not highly vulnerable.

Blighted leaves will turn brown and die but remain attached to twigs. The twig may curl downwards producing a "shepherd's hook", and may release a clear amber liquid. Symptoms usually appear in spring caused by the bacterium *Erwinia amylovora*. Bacteria can spread by wind, rain, insects and contaminated pruning tools.

Control: Prune back infected branches to 25cm past the diseased wood. Branches should also be burned or buried immediately. Pruning tools should be dipped in Lysol or bleach after each cut is made to avoid the spread of infection.

Harvesting

Harvesting plums depends on the maturity time of the cultivar. It can also be a matter of personal taste. The timeline for harvesting is generally between mid-August and late September. If plums are to be shipped to market, they should be picked slightly earlier and left unwashed. Fresh plums may be left on tree until ready for home usage. They are suitable for eating fresh, making pies, sauces, jellies or any recipe calling for plums.

Plum trees will begin to produce fruit at four or five years old. Like most fruit, leave the plums on the tree until they are ripe and ready to eat. Ripe plums will be produced over a month or so and it is best to pick a few but often over this time. Any diseased fruit should be removed immediately because they will only attract diseases and wasps.

Ripe plums keep for only a few days and are best stored in the fridge. If you want to keep plums for longer, pick them when they are slightly under-ripe, placing them in a paper-lined box in a cool dark place.

Recommended Cultivars

There has been extensive research and breeding that has gone into producing winter hardy plum species for the prairie climate. The following is a list of University of Saskatchewan recommended cultivars:

Dessert Plums

- *Denniston's Superb:* An early variety of plum, with medium-size green fruits flushed with red. These plums have an excellent, sweet flavour.
- *Gaviota*: These large round plums have yellow skins deeply tinged with scarlet, and sweet, juicy, red flesh. They are best eaten raw.
- *Marjorie's Seedling:* These small purple plums with a green flush have bitter skins and sweet, green, almost translucent flesh. They are good for eating and cooking.
- *Santa Rosa and Burbank:* Large and round, with bright red skins, these two North American varieties of plum are mainly grown in California. They have juicy, deep yellow flesh and a pleasantly tart flavour, which makes them good for both cooking and eating.
- *Victoria:* The most prolific of all dessert plums, Victorias were first cultivated in 1840 from a stray seedling found in Sussex, England. Since then, these large oval fruits with yellow skins, flushed with scarlet, and sweet, juicy flesh have become ubiquitous. They are good for bottling and canning, stewing or eating raw.

Cooking Plums

- *Beach plums:* These small plums grow wild along the Atlantic coast of North America, especially near Cape Cod. They have dark purplish-black skins and tart flesh which makes them unsuitable for eating raw, but they make excellent jams and jellies.
- *Cherry plums or mirabelles:* These very small wild plums are round and grow on long stalks like cherries. They have black, red or yellow skins, which can taste rather bitter, but all have sweet, juicy flesh. These plums can be eaten raw, but are best stewed or baked, or made into jams, sauces and jellies.
- *Golden mirabelles:* These plums are delicious in tarts and souffles, and are also made into a plum eau-de-vie, called mirabelle.
- *Czar large:* This species is a rather acidic dark blue-black plums with golden flesh, these can be eaten raw, but are more usually used for cooking. They are best eaten straight from the tree.
- *Quetsch:* Also known as svetsch or Zwetschen, these small purplish-black plums have a beautiful bloom. Although their flesh is sweet, they are seldom eaten raw, but are used in Eastern Europe to make plum breads and pflaumenkuchen, yeast dough topped with purple plums. Quetsch plums are also used for making slivovitz and other plum brandies.

USES OF PLUMS

Plums can be used in a wide variety of recipes - check out the plum recipe page for some delicious suggestions.

However, their uses do not end there. The laxative effects of the fruit both fresh and dried as prunes is very well known. But what is perhaps less well known is the cooling lubricant properties of the fresh flesh, although we do not recommend rubbing them on any wounds!

Nutritional value

Plums contain more antioxidant than any other fruit. They provide about 40 kilocalories per 100g/33/40Z.

Buying and Storing

Plums are delicate, so make sure that the ones you buy are unblemished. They should be plump and firm, with some "give" (but never squashy), and they should be fully coloured for their variety. Plums should always have a pleasant aroma. These fruit ripen fast and quickly become over-ripe, so store them in the fridge for only a day or two. For cooking at a later date, plums can be frozen: halve the fruit and remove the stones. Place on trays and open freeze, then pack the fruit into polythene bags and seal.

Preparing and Cooking

Dessert plums are delicious eaten on their own. Dual varieties (suitable for eating and cooking) and cooking plums make excellent pies and tarts, compotes, crumbles, dumplings, sauces, mousses and souffles. They can be poached, baked or stewed, either whole or in halves or slices. It is not recommended that plums with tough skins are cooked in the microwave, as they will not soften in the short cooking time. Cook plums until just tender; do not let them disintegrate. Plums make tasty ice cream, and the poached fruit goes very well with ice creams flavoured with spices like cardamom, nutmeg and cinnamon.

Plums go extremely well in savoury dishes. The Chinese make them into a thick sweet-sour sauce to serve with Peking duck, lamb or pork. Spiced stewed plums are good with gammon, cured meats, terrines and poultry. Plums add a special flavour to beef or lamb casseroles.

Plums also make superb jams and jellies. They can be preserved in many different ways; dried (as prunes), crystallised and candied, bottled or made into wine and liqueurs like slivovitz and plum brandy.

REFERENCES

Galán Saúco, Víctor. "Current Situation, Trends, and Future of Agronomic Research on Tropical Fruits." In *Proceedings of the International Conference on Tropical Fruits, Kuala Lumpur, Malaysia, 23–26 July 1996.*

Galleta, Gene J., and David G. Himelrick, (eds.) 1990. *Small Fruit Management*. Englewood Cliffs, N.J.: Prentice Ha

Jason Zasky. "Turning Over a New Leaf Change from 'Prune' to 'Dried Plum' Proving Fruitful". http://www.failuremag.com/arch_business_dried_plums.html. Retrieved on 2008-01-26.

9

Strawberries

Garden strawberries are a common variety of strawberry cultivated worldwide. Like other species of *Fragaria* (strawberries), it belongs to the family *Rosaceae*. Technically, it is not a fruit but a false fruit, meaning the fleshy part is derived not from the plant's ovaries (achenes) but from the peg at the bottom of the bowl-shaped hypanthium that holds the ovaries. The Garden Strawberry was first bred in Europe in the early 18th century via an accidental cross of *Fragaria virginiana* from eastern North America, which was noted for its flavor, and *Fragaria chiloensis* from Chile, which was noted for its large size. Cultivars of *Fragaria × ananassa* have replaced in commercial production the Woodland Strawberry, which was the first strawberry species cultivated in the early 17th century.

The name strawberry is believed to have been derived from the name strewn berry. Berries that are strewn among the leaves of the plants were known as strewn berries in the olden days. When time passed by, the term strewn berry gradually came to be pronounced as strawberry by the people. It also says that the name "straw" berry comes from the practice of the farmers making mulching over the plants of strawberries by using straw.

Strawberries belong to the Rosaceae family. The typical strawberries that we use today belong to the genus Fragaria. The name fragaria (odorous) indicates the nature of the perfumed flesh of strawberries. It says Madam Tallien, a great figure during the French revolution used to take bath using strawberries to keep her skin radiant, smooth and free of blemishes. Fontenelle, a centenarian writer of the 18th century gave all credit for his long life to his habit of eating fresh strawberries every day. Ironically the same strawberries were considered as poisonous in Argentina till the mid of nineteenth century.

Strawberry, against the belief of many of us, is not a fruit but on the other hand an enlarged receptacle of the flower of the plant. Strawberries have a history of more than

2200 years. Wild strawberry plants were found in various parts of Italy as long ago as in 234 BC. The first time these wild strawberries were discovered by some Europeans when their ship landed in the shores of Virginia in 1588. History says that early settlers in Massachusetts were very much fond of the strawberries grown by the local American Indians who started strawberry cultivation as early as 1643. After 1860 strawberries were grown in many parts of the country on a large scale.

Figure 1. Strawberries

The history of strawberry cultivation in California begins from the year 1900. Now it has become the largest supplier of quality strawberries that satisfies almost eighty percent of the strawberry requirements of the nation. Almost 25000 acres in different parts of California is now utilised for the purpose of strawberry cultivation.

Strawberries that were originally grown in northern parts of Europe were also found in different parts of Russia, Chile and United States. Strawberries are also thought to have been cultivated in ancient Rome in large scale for medicinal purposes. During the 13th century, France also cultivated strawberries for its medicinal purpose. The roots, fruits and leaves of Alpine strawberry were widely used in these countries to make medicine that were used to cure digestive problems and skin diseases. Berry was used internally to cure diarrhea. Its leaves and roots were used for gout. Over the years strawberries were used for treating skin rashes, sun burn, discoloration of teeth and digestive disorders. In 13th century, strawberry was widely used as an aphrodisiac.

During the medieval period strawberry was considered as a symbol of prosperity and richness, peace and perfection in the society. And for this reason strawberries were served lavishly while conducting various events under state control.

The modern strawberries that we use today belong to the genus Fragaria. The fruits that we now consume come from the Americans, a hybrid variety of the strawberries seen in North and South America. It belongs to the class of Magnoliopsida under the division of Magnoliophyta. Rosaceae is the scientific family name of strawberries. The sub family to which this fruit belongs to is known as Rosoideae. There are more than twenty species of named varieties of strawberries found on earth till today. A large variety of hybrids of this fruit is also now available in the market.

The most common varieties of this fruit are grown commercially and belong to the cultivators of Garden Strawberry. The taste of a strawberry fruit basically depends on the cultivator. Strawberry is considered as an accessory fruit. Unlike other fruits the fleshy parts of this fruit is not derived from the ovaries but from the peg at the bottom of the hypanthia that are attached to the ovaries of the plant. For this reason the fleshy part of the strawberry cannot be called as the fruit of the plant. It is the seeds that we find on the outer part of the strawberry constitute the real fruit of the plant.

The typical strawberries that we now see in the market actually come from the Americans who have dared to experiment them with two different varieties of strawberries found in North and South America. This new hybrid variety of strawberries was nothing but an outcome of the attempt made by the European horticulturalists to rectify a mistake of taking only female strawberry plants from South America to their labs. They had no other option but to cross them with the North American varieties to get fruits and seeds from the plants that they had brought from South America.

The genus name of strawberry; Fragaria comes from the term "fragans" indicating the perfumed nature of the flesh of strawberry. This fruit is known for its ability to cure diseases such as digestive disorders and gout. It can also be use as an effective tool to keep the radiance of the skin by making the skin free of blemishes, rashes and for protecting from sunburns. Madam Taliien, a known figure during the French revolution used to take strawberry baths to keep her skin fresh, radiant and free from blemishes. Until the mid of nineteenth century strawberries were considered as poisonous in Argentina.

There exist two versions about the origin of the name strawberry. According to the popular etymology, it is the practice of the cultivators who used to make mulch using straw to protect fruits from rotting hence the name 'straw' berry to this fruit. The other version regarding the origin of the name strawberry believes that the name strawberry come from the strewn nature of the plant that spreads around the ground. The preachers of this version believe it to be originated from the Anglo-Saxon verb for 'strew" meaning to spread around. However there exists no evidence to show whether Anglo-Saxons ever grew strawberries in their land

CULTIVATION

Strawberry cultivars vary remarkably in size, color, flavor, shape, degree of fertility, season of ripening, liability to disease and constitution of plant. Some vary in foliage, and some vary materially in the relative development of their sexual organs. In most cases the flowers appear hermaphroditic in structure, but function as either male or female.

Figure 2. Garden Strawberry plant

For purposes of commercial production, plants are propagated from runners and generally distributed as either bare root plants or plugs. Cultivation follows one of two general models, annual plasticulture or a perennial system of matted rows or mounds. A small amount of strawberries are also produced in greenhouses during the off season.

The bulk of modern commercial production uses the plasticulture system. In this method, raised beds are formed each year, fumigated, and covered with plastic, which prevents weed growth and erosion, under which is run irrigation tubing. Plants, usually obtained from northern nurseries, are planted through holes punched in this covering. Runners are removed from the plants as they appear, to encourage the plants to put most of their energy into fruit development. At the end of the harvest season, the plastic is removed and the plants are plowed into the ground. Because strawberry plants more than a year or two old begin to decline in productivity and fruit quality, this system of replacing the plants each year allows for improved yields and denser plantings. However, because it requires a longer growing season to allow for establishment of the plants each year, and because of the increased costs in terms of forming and covering the mounds and purchasing plants each year, it is not always practical in all areas.

The other major method, which uses the same plants from year to year growing in rows or on mounds, is most common is colder climates. It has lower investment costs, and lower overall maintenance requirements. Yields are typically lower than in plasticulture.

Strawberries are often grouped according to their flowering habit. Traditionally, this has consisted of a division between "June-bearing" strawberries, which bear their fruit in the early summer and "Ever-bearing" strawberries, which often bear several crops of fruit throughout the season. More recently, research has shown that strawberries actually occur in three basic flowering habits: short day, long day, and day neutral. These refer to the day length sensitivity of the plant and the type of photoperiod which induces flower formation. Day neutral cultivars produce flowers regardless of the photoperiod.

Strawberries may also be propagated by seed, though this is primarily a hobby activity, and is not widely practiced commercially. A few seed-propaged cultivars have been developed for home use, and research into growing from seed commercially is ongoing. Seeds are acquired either via commercial seed suppliers, or by collecting and saving them from the fruit.Most strawberries are now fertilised with artificial fertilisers, both before and after harvesting, and often before planting in plasticulture. The harvesting and cleaning process has not changed substantially over time. The delicate strawberries are still harvested by hand. Grading and packing often occurs in the field, rather than in a processing facility. In large operations, strawberries are cleaned by means of water streams and shaking conveyor belts.

Varieties

A large number of varieties are available. For the hilly areas, varieties Royal Sovereign, Srinagar and Dilpasand are suitable. Some of the introductions from California, such as Torrey, Toiga and Solana may prove even more successful. The variety found successful in Bangalore has been named Bangalore and which has performed well at Mahabaleshwar also. For the north Indian plains, Pusa Early Dwarf which has dwarf plants, large firm wedge-shaped fruits, has been recommended. Another variety with rich aroma but softer fruits is Katrain Sweet. Some of the varieties found successful in warmer parts of the U.S.A. are: Premier Florida-90, Missionary, Blackmore, Klonmore & Klondike. Some of these may prove successful for cultivation in Indian plains.

Propagation

Propagation is done by means of runners that are formed after the blooming season. The plants may be allowed to set as many runners as possible but not allowed to set any fruits. All the plants with good root system should be utilised to set a new plantation. Given the best attention and care, a single plant usually produces 12 to 18 runners.

Choosing Strawberry Cultivars

Strawberry plants make an excellent addition to the home garden. 30 plants will provide enough strawberries for a family of four. The attractive plants can be grown as a groundcover or landscape ornamental, or in a traditional bed. Choose vigorous, "virus free" plants from an established nursery. Buying from a local nursery ensures that the cultivar you have chosen will do well in your area, but mail-order nurseries will offer the widest selection.

Three types of strawberries are readily available to the home gardener:

- June Bearing strawberries produce a large, concentrated crop in late spring. You'll get more berries per season from June bearing strawberries, and the entire crop will ripen over a three week period.
- Ever Bearing strawberries produce one crop in spring and another in fall.
- Day Neutral plants are capable of producing fruit throughout most of the growing season.

Site Selection and Preparation

Strawberries need full sun to produce the best crop, but you can get a harvestable crop with as little as 6 hours of direct sunlight per day. Increasing the amount of sunlight that your strawberries receive will result in more and better quality berries.

Strawberries will not tolerate drought or standing water. Be sure that the site you select is well-drained. Add organic material if the soil does not retain moisture well. Beds should be placed beyond the root zone of large trees so that the tree will not compete with the strawberries for moisture. Choose a location that will be easy to water during dry spells.

A pH of 5.3 to 6.5 is optimal, but strawberries will grow in soils with a pH between 5.0 and 7.0. Test kits are available in most garden supply stores, but if you plan to do a lot of gardening, a professional soil test is highly recommended. Soil tests are available through most county agricultural extension agents.

The site selected should be free from weeds, grubs and soil-borne diseases. Areas where sod has grown should be tilled and cultivated the year before planting to eliminate grass that will compete with the strawberries. Perennial weeds also compete with the plants and cause a marked reduction in production. Strawberries are susceptible to Verticillium Wilt, and should not be planted where tomatoes, peppers, eggplant and potatoes have grown in the past three years since these vegetables often carry the disease. If you must plant where these vegetables have grown, choose resistant varieties of strawberries such as Allstar, Earliglow, Settler, Tribute or Tristar.

Planting

The land for strawberry planting should be thoroughly prepared by deep ploughing followed by harrowing. Liberal quantities of organic manure should be incorporated in the soil before plating. Strawberry can be planted on flat beds, in the form of hill rows or matted rows, or it can be planted on raised beds. In irrigated areas, plantings on ridges is advised. In Mahabaleshwar, the usual practice is to plant on raised beds 4 x 3 meters or 4 x 4 meters. The planting distance should be 45 cm from plant to plant and 60 to 75 cm. from row to row. In the hills, Transplanting is done in March-April, September-October, but in the plains, the months of January-February may be utilised for this purpose. At Mahabaleshwar normally strawberry is planted during November-December.

The plants should be set in the soil with their roots going straight down. The soil around the plant should be firmly packed to exclude air. The growing point of the plant should be just above the soil surface. During planting, the plants should not be allowed to dry out and should be irrigated immediately after planting.

It is pretty easy to grow strawberries in gardens and in containers at home. Before you start to grow strawberries you have to acquire minimum knowledge about the three distinct types of strawberry plants available in nurseries. These plants include summer bearing, perpetual bearing or ever bearing and alpine.

Perpetual bearing strawberries will bear fruits through out the year although the bulk of the fruits will come only in the season that extent from April to October. You can expect fruits from summer bearing strawberries for more than three weeks a year.

Select a site where you can assure maximum sun to grow strawberries. It should also be a place where the soil is well drained. If you are planning to grow strawberries in your garden where you grow other plants do not choose an area where you have grown potatoes to avoid virus infection to the spuds. Long days and warm temperature is the best and the most suitable climate that favor maximum growth of the leaves and runners of a strawberry plant. Short days and cool temperature is good for flower formation in short day varieties of strawberries. You can expect tastier fruits from growing strawberries in areas where the days are warm and sunny and the nights are cool.

If your plan is to grow strawberries in the ground then you should prepare raised bed to plant them. The highest ground is ideal location to place strawberries and it should always be protected from the frost by placing them at an ideal location. It is better to choose a location where plants that belong to the Solanaceae family were not grown for at least three years.

Prepare the ground by mixing it with compost or well composted manure. Vegetable food fertilizers can also be used to prepare land for growing strawberries. Once the preparation of the land is over, you can cover the land by using a weed block material. Mark the places where the strawberries are to be planted by cutting an X shape in the material with which you have covered the soil. You will be able to protect the berries from the attack of snails, earwigs, slugs and saw bugs by covering the land with the weed block material. It is advisable to prepare the land in this way at least one month before you actually plant the strawberries.

The beds prepared for growing strawberries should have at least be five to six inches height and six to twelve inches width. Arrange the plants eight to fourteen inches apart in a single row on the top of the bed that you have prepared to plant the strawberries.

Any method of plantation that allows the root of the plant to go straight down and spreading dove-tail can be adopted for planting strawberries. Make sure that the soil is firmly pressed against the roots when you plant the strawberry plant in the ground or in the container. The crown of the plant should be above the surface covering all the roots.

Strawberry plants with shallow roots require watering through the growing season. Drip irrigation is the best method of irrigation to be used in a strawberry garden.

Strawberries are usually grown using either the matted row or the hill system. The matted row system requires the least maintenance.

Plants grown in an ornamental setting can be incorporated into the landscape. Day neutral plants, which produce fewer runners, work well in the front of a border and along sidewalks and driveways. Alpine strawberries produce tiny, highly flavored strawberries and make excellent ornamentals. To use strawberries as a groundcover, space the mother plants evenly 1 to 2 feet apart. Weed well the first year, and thereafter maintenance should be minimal. Strawberries grown as a groundcover will not produce much fruit. In all cases, the soil should be well tilled and fertilised a week before planting.

Matted Row System

The matted row system works best with June bearing strawberries. You'll need an area about eight feet wide and 30 feet long to accommodate 30 plants. In the spring, plant the original (mother) plants 2 feet apart in rows 3-4 feet apart and allow them to produce and set runner (daughter) plants freely. All the plant's energy is devoted to producing and developing strong plants the first year, so all flowers should be picked off. No fruit is harvested until the second year. Keep rows to a width of 18 to 24 inches. If plants are vigorous you will have to cut back runners that grow into the pathway between the rows.

Hill System

This system is used for day neutral or everbearing cultivars and works well in raised beds. Rows should be about 8 inches high and 26 inches wide. Mother plants are planted only 12 inches apart, and staggered double rows are preferred. Leave an isle between the rows about 2 feet wide.

The plant's energy is put toward producing berries rather than producing an abundance of plants as in the matted row system, so all runners are removed and berries are harvested the first year. Plants become less productive over time, so they should be replaced every 1-3 years.

The hill system works well in warm climates where strawberries can be planted in the fall. The plants can be treated as annuals, replacing them every fall for best results.

Setting the Plants

Strawberries are planted in the spring as soon as the soil can be worked. If plants must be stored before they are planted, choose a cool location out of direct sunlight and keep the roots moist, but not soggy. If possible, plant on a cool, cloudy day to reduce the stress on the transplants. Follow these guidelines to prepare your plants:

— Prune damaged roots

— Trim excessively long roots to 4-5 inches in length.

— Remove all flowers, runners and old leaves

— Place the transplants in a container with a small amount of water in the bottom and keep them out of direct sunlight.

Plants should be set with the roots pointing downward and forming a small fan. If plants are set too shallow the crown may dry out, and if too deep the crown may rot. Set the plants just deep enough so the midpoint of the crown is even with the soil surface. After setting in, firm the soil around the plant and water thoroughly.

Care of young Plantation

The roots of strawberry plants spread out close to the surface. Therefore, the soil should be well supplied with moisture, and hoeing should be done lightly and young plantation be kept weed free.

Special Horticultural practices

In cold climate the soil is covered with a mulch in winter to protect the roots from cold injury. The mulch keeps the fruits free from soil, reduces decay of fruits, conserves soil moisture, lowers soil temperature in hot weather, protects flowers from frost in mild climates and protects plants from freezing injury in cold climates. Several kinds of mulches are used, but the commonest one is straw mulch. The name strawberry has been derived from this fact. Black alkathine mulch is also used to cover the soil. It saves irrigation water, prevents the growth of weeds and keep the soil temperature high.

Irrigation

Since strawberry is relatively shallow-rooted, it is susceptible to conditions of drought. Planting early in autumn allows the plants to make good vegetative growth before the onset of winter. However, in this case it is necessary to ensure that newly planted runners are irrigated frequently after planting, otherwise the mortality of the plants becomes high. During September and October, irrigation should be given twice a week if there is no rain. It may be reduced to weekly intervals during November. In December and January, irrigation may be given once every fortnight. When fruiting starts, the irrigation frequency may should again be increased. At this stage frequent irrigation gives larger fruits.

Application of manures and fertilisers

Strawberry requires moderate amounts of nitrogen. Addition of organic matter to the soil, in the form of 50 tons of Farm Yard manure per hectare is highly desirable. It improves the water holding capacity of the soil and also gives better runner formation. Farm yard manure may be supplemented by chemical fertilisers to make up the total quantity of nitrogen from 84 to 112 kg per hectare, Phosphorus 56 to 84 kg per hectare, and Potash 56 to 112 kg per hectare. The Phosphatic fertiliser should be incorporated into the soil before plantings. The nitrogenous fertiliser be applied in Two doses and potash at the time of flowering only. Application of adequate amounts of nitrogen gives higher yield of early berries.

Plant protection

Red spider mites and cutworms are important pests of strawberry. The mites can be controlled with 0.05 per cent Monocrotophos + 0.25 per cent wetable sulphur. The cut worms can be controlled by dusting the soil before planting with 5 per cent chloradane or Heptachlor dust at the rate of 50 kg per hectare and mixing it thoroughly in the soil by cultivator.

The two commonest diseases of strawberry are red stele, caused by the fungus Phytophthora fragariae and black root rot. The remedy for the former lies by growing resistant varieties like stelemaster and for the latter to maintain the vigour of the plants and rotate strawberry with other crops like legume vegetables (beans, peas etc). Strawberry also suffers from virus diseases known as yellow edge, crinkle and dwarf. Raising of strawberry nursery in the hills helps to check these. Strawberry also throws some chlorotic plants, which result from genetic segregation. These should not be confused with virus affected plants and should be rogued out.

Harvesting and Yields

The fruit ripens during late February to April in the plains and during May and June at high elevations like Mahabaleshwar, Nainital and Kashmir. For local market the fruit should be harvested when fully ripe, but for transport to distant markets, it should be harvested when still firm and before colour has developed fully all over the fruit. Harvesting should be done preferably daily. Since fruit is highly perishable, it is packed in flat shallow containers of various types (cardboard, bamboo, paper trays etc.) with one or two layers of fruits. Harvesting should be done early in the morning in dry conditions. Washing the fruit bruises it and spoils its lustre.

The yield varies according to season and locality. A yield of 20 to 25 tons per hectare is excellent, though yields upto 50 tons per hectare have been reported under ideal conditions.

Picking Strawberries

Strawberries look better and keep longer when they are picked and handled correctly. Because they are a very tender fruit, they will bruise and discolor any time they are squeezed, so handle them gently.

Picking pointers:

1. Grasp the stem just above the berry between the fore-finger an the thumbnail and pull with a slight twisting motion.
2. With the stem broken about 1/2 inch from the berry, allow it to roll into the palm of your hand.
3. Repeat these operations using both hands until each holds 3-4 berries.
4. Carefully place — don't throw — the fruit into your containers.
5. Don't overfill your containers or try to pack down the berries.

Whether you pick strawberries from your own garden or at a Pick-Your-Own farm, here are a few tips to keep in mind:

- Be careful that your feet and knees don't damage plants or fruit in or along the edge of a row.
- Remember that heaping strawberries more than 5 inches deep will bruise the berries.

- Only pick the berries that are fully red. Part the leaves with your hands to look for hidden berries.
- Remove berries that show rot, sunburn, insect injury, or other defects and destroy them.
- Pick berries for immediate use any time but if you plan to hold the fruit for a few days, try to pick in the early morning or on cool, cloudy days. Berries picked during the heat of the day bruise easily and will not keep well.
- Keep picked berries in the shade and cool them as soon as possible after picking. Berries that have been handled carefully will keep up to three days in the refrigerator.

Strawberries are highly perishable and hence a great deal of care in harvesting and handling as well as its marketing also requires to be organised carefully. Usually the fruit is picked in the early morning and sent to the market in the afternoon of the same day or is picked in the late afternoon, stored overnight in a cool place, and sent to market the following morning.

TECHNIQUES FOR STRAWBERRY CULTIVATION

Different varieties of strawberries differ in taste, color and size and are now available in the market. Most of the strawberries that we consume are the ones produced through cross methods of horticulture. Each of these plant varieties also differ in their degree of fertility, season of ripening, liability to get infected by diseases and in the constitution of the plant.

To produce strawberries on a large scale commercial basis, the plants are normally collected from runners. These baby plants will normally be distributed among the cultivators as bare root plants or as plugs. The two methods of cultivations adopted for commercial production of strawberry include annual plasticulture or a perennial system of matted rows. During off season a small amount of strawberries are produces following the green house system of cultivation.

Still the bulk cultivation of strawberries for commercial uses follows the plasticulture system of cultivation. In this method, strawberries will be planted on a raised mount of soil known as beds. The beds prepared to plant strawberries will be covered by using a black plastic known as weed blockers, to prevent weeds and erosion from the farm. Irrigation tubing system will run under these weed blockers.

The strawberry plants will then be planted through the hole marked on the weed blocker placed on the strawberry bed. Runners that form near to the plants will be removed to encourage the plants to use more energy into fruit development and to get the maximum from the cultivation.

At the end of the season, the weed blocker will be removed and the plants will be plowed into the ground. A plant more than one or two years old will not be able to produce as much fruits as a new plant in the beginning. Replacing of plants every year will allow the cultivator to obtain maximum fruits from the cultivation. Though desirable, considering the high cost involved in preparing the soil for planting the plants, many do not attempt to follow this method of replanting strawberry plants every other year to enjoy more fruits out of the cultivation.

To get the best out of the plantation it is best to renew the plantation every second or third year. Plantations that are kept free from runners need not always replace the plants like the cultivations that allow runners to grow with the main plants. The time for re plantation can be extended by cutting off the leaves of the plants after the fruits have been collected from them. After cutting of the leaves a good top-dressing of rotten dung or leaf mould should be applied on the plant to keep them in the plantation for a longer period.

A mulching of straw manure placed between the plants during spring is desirable to keep the fruit clean and the ground moist. The plants should be watered regularly during the dry weather after the fruit is set on the plant. Once the fruit season is over the runners will be removed from the ground to keep the ground loosened and raked.

Growing Hydroponic Strawberries

Hydroponics is defined as the science involved in the growth of plants without the need of soil. The nutrients gained by the plants from soil are included in water. The word originated from the Greek words hydro which means water and ponos meaning labor. The first use of hydroponics was reported during the Hanging Gardens of Babylon, one of the old Wonders of the World. It is said that the plant growth happened along the stream of water. The historic evidence speaks that the soldiers stationed in Pacific Islands consumed fruits and vegetables generated by hydroponics.

You might be wondering the reason why hydroponics is utilised for strawberries. There is a valid motive behind this aspect which is to protect the environment from an alarming problem. Methyl bromide was the chemical on which the strawberry growers depended to keep away soil borne pests, insects, weeds and diseases. This kind of fumigation had its own significance as it helps in yielding top quality strawberries. Presently, it has been found that methyl bromide is a major threat to the ozone layer and for this reason the chemical is banned all over. Due to this reason, the strawberry growers have thought of switching to hydroponics to make it environment friendly and also increase the production as a whole.

Hydroponic strawberries are making a good impression and a good alternative to cultivators. Cultivators across the globe are switching their cultivation style to hydroponics as it has been recognised to be one of the most productive and efficient forms of production.

Hydroponic strawberries does add a few advantages for the cultivators, the most important being the remarkable taste which just keeps you yearning for more. Since the cultivation of these plants is at an elevated height, the picking rate has been increased with no fatigue and a much easier way of cultivation.

When considering the yields per plant, it is much higher and the losses are lower when compared to the growth in soil. You can do the cultivation on any land conditions be it poor or rich because the problem of weeds is simply eliminated. Vertically tying the garden could utilise more space and can be extended sky high thus stacking to utilise more space. Hydroponic strawberries are not seasonal therefore you can cultivate them all year round.

Hydroponic cultivation can be even done right at home which could feed a family of four easily. The cultivation could be done either indoor or outdoor with the help of artificial light or sunlight. Certain steps should be taken into consideration before you start up with the cultivation process right at home. Care is to be taken while you buy the cold-stored runners. It should be purchased only from a well known local nursery after being duly certified virus tested. If the steps are being followed appropriately, you can get a higher yield of these succulent fruits right at your home.

Growing Strawberries in Containers

Strawberry plants require special climatic conditions such as full day light and a well drained site to grow healthy and to obtain maximum yield. It also requires a place where you have not been planting vegetables like potatoes, tomatoes, egg plants or other similar varieties of vegetable crops. This is to ensure that your strawberry plants do not get infected by any form of virus. If no such area is available in your garden, you can go for container growing of strawberries.

Growing strawberries in a containers does not require much effort to start with. It can be done as easily as you do any other container gardening where you plant plants in a container or in a pot. You can either select those pots that are specially designed for growing strawberries or a can or a container to grow strawberries. You can easily transport these containers to the places where the plant can get full sunlight which is an essential requirement for the healthy growth of a strawberry plant. It can also be placed anywhere in your yard or even on your windowsills conveniently.

To grow strawberries in containers you should have strawberry crowns, preferably the crowns that are certified as disease free, pots or cans, potting mix, fibre glass window screen and slow release fertiliser. If available it is always advisable to go for the pots that are designed to grow strawberries if you wish to grow strawberries in a container. These pots are available in different materials such as plastic, terracotta and in ceramic. Though a little expensive than the other two varieties, ceramic strawberry pots are the best choice to start with.

If no such pots are available near your place of residence, you can even do the container gardening of strawberries by using cans that are available at your home. Narrow planter boxes with an approximate depth of six to eight inches and seven inches wide can be used for planting strawberry crowns. When you use containers or cans for the purpose of planting strawberries, make nail holes on the bottom of the container or the can to avoid water storage inside the can when you water the plants in the container. Cover the holes that you have made on the bottom of the can using a fibre glass window screen to prevent erosion of soil through these holes.

The pots that are designed for pot planting of strawberries will always have pockets up and down in the sides of the pot. Choose one plant for each pocket of the pot and three or four plants to be planted on the top. You can use all purpose soil mix to fill the container when you plant strawberry crowns in them. If you wish you can also add some reliable slow-releasing fertiliser in the ready made pot mix that you use to fill the pot. You can avoid subsequent fertilisation of plant by mixing slow-releasing fertiliser with the potting soil.

Growing strawberries in containers is especially flexible for moving the plants around whenever you want and for relocating the plants to different places. Take advantage of this flexibility with strawberry growing hanging baskets or plastic pockets for a very creative gardening solution.

PESTS

Around 200 species of pests are known to attack strawberries both directly and indirectly. These pests include slugs, moths, fruit flies, chafers, Strawberry Root Weevils, Strawberry Thrips, Strawberry Sap Beetles, the Strawberry Crown Moth, mites, aphids, and others.

Strawberry Bud weevil

Adult is a snout beetle about 1/10 inch (3 mm) long, chestnut brown in color, and possessing two black spots on its back (elytra). Larvae are tiny creamy white-colored grubs found inside unopened flower buds.

The strawberry bud weevil is probably one of the most important direct pest of strawberries in the United States. This pest has been shown to cause yield losses from 50 to 100% in some areas. The strawberry bud weevil is widespead and occurs throughout virtually all the strawberry-growing regions of this country, including Virginia. Strawberry bud weevil has one generation per year. Overwintering adults emerge early in the season from ground litter commonly in wooded areas and migrate to strawberry fields (around late April in the Mid-Atlantic region). Ovipositing females puncture unopened buds with their long beaks and deposit a single egg into the bud. Damage results when females sever the strawberry bud from the pedicel following oviposition, causing it to hang by part of the stem, or fall to the ground, thus, preventing fruit formation. Larvae develop in the severed buds and reach maturity in 3-4 weeks. Adults emerge in June, feed on flower pollen, then enter an estavation in mid-summer and remain inactive the rest of the season.

Strawberry bud weevil (Clipper)

One method of field scouting involves sampling weevils on plants, during the early blossom/bud stage. Shaefers suggests an economic injury level of 1 female beetle per 40 row feet. Cooley & Schloemann developed a sampling program using bud damage as a sampling index. According to this program, sample units of 2 row feet of plants are sampled at 5 to 10 locations along a v-shaped transect. Treatment is advised if an average of 0.6 clipped buds per row foot are found.

Recent research in New York indicates that only rows near block edges near woods may need to be sprayed. If compensatory varieties such as Seneca are used, sprays may not be needed at all.

Pistillate varieties of strawberries are relatively immune from attack since only varieties with staminate flowers seem to provide adequate food for developing larvae. Early-fruiting varieties are more susceptible to attack than later-fruiting varieties. Schaefers recommends topping of plants and removal of foliage and mulch immediately following harvest, then applying a follow up chemical spray to kill overwintering adults. Other cultural practices include avoiding field site selection near wooded areas to prevent high numbers of overwintering adults from entering the field in the spring. Mulches and full canopy beds encourage adults to overwinter and remain in the field, thus plowing of old beds immediately following harvest causes adult mortality. Cropping fields less than three years is also recommended.

Lygus bug/Tarnished Plant Bug

Adults are about 6 mm long, oval, and rather flattened. They are yellowish to greenish brown, and have dark markings on their wings. The immature forms are pale green and look similar to an aphid. They can be distinguished from aphids by their more rapid movements.

The piercing-sucking feeding by Lygus bugs are one of the causes of irregularly-shaped, catfaced strawberries. The insects damage fruit by puncturing individual seeds; this, in turn, stops development of the berry in the area surrounding the feeding site. Straw brown seeds that are large and hollow are a good indication of lygus bug damage. There are 3 to 6 generations of this pest each year. Adults become active in early spring and deposit their curved eggs into stems, and leaf midribs. Egg hatch takes place about 1 week later depending on temperature, and the green-yellow nymphs progress through 5 instars, reaching the adult stage in approximately 30 days.

One method of field scouting involves shaking flower clusters over a sheet of white cardboard. An action threshold of 1 nymph per cluster is recommended. Penn suggest a threshold of 20-25 TPB per 50 clusters. Other scouting methods have been implemented in California. Threshold levels have been developed for lygus bugs depending on monitoring method used. When the beat sheet (12 inch embroidery hoop with muslin) is used, sample one plant in each 20 ft of row by placing the beating tray under the plant and beating it with your hand. Apply sprays when 1 lygus nymph is found in 20 plants. The Allen-vac is a more efficient sampling device, and sprays should be applied when you find 1 lygus/10 plants.

Strawberry Sap Beetle

Sap beetles often fly to ripening or damaged raspberries, strawberries, melons, early apples, tree wounds, corn, tomatoes, and osage orange fruits. They may bore into the fruit, eat a portion and make it unfit for human consumption. The picnic beetle is essentially a secondary invader of damaged plants and decomposing plant tissue, but in undamaged ear corn silks and ripe raspberries, it can be a primary invader. The strawberry sap beetle is a primary invader of ripe and nearly ripe strawberries. As a nuisance, sap beetles may congregate in annoying numbers on screen doors, around garbage cans, invade homes, backyards, picnic areas, food processing plants, and roadside fruit and vegetable stands.

One of the identifying characteristics of all sap beetles is their "knobbed" antennae.

Picnic beetle adults are about 1/4-inch long and black with four orange-red spots on the wing covers. Eggs are milky-white, sausage-shaped, and about 1/32-inch long.

Dusky sap beetle adults are about 1/8-inch long with short wing covers and are uniform dull black in color.

Strawberry sap beetle adults are slightly less than 1/8-inch long, light to dark brown, oval, and somewhat flattened. Larvae of all three are white. Pupae are white, turning cream-colored and later tan before adult emergence.

All of the sap beetles in Ohio overwinter as adults in protected places such as decaying vegetation, debris or fruit buried in the ground. In the spring, picnic beetle adults come out of hibernation and mate. Egg laying begins in April and continues in May and June. Females lay 5 to 15 eggs per day, scattered at random near decomposing plant material rather than on the material itself. Larvae develop in spilled grain, feed, corn ears, waste onion piles, and soil saturated with juices and food material in contact with the soil. Full-grown larvae leave their food when mature, wander through the soil and change to the pupa stage. Adults then emerge in June and July. The cycle from egg to adult takes about 30 to 35 days. There is usually only one generation per year.

Newly emerged adults do not lay eggs but congregate on screen doors, around garbage cans, in picnic areas and parks, and about anywhere food is grown or being served. They are a general nuisance, attracted to sweet or fermented plant juices. Beetles are found on cracked tomatoes, damaged sweet corn ears, overripe muskmelons, strawberries, and raspberries.

The life cycle of the dusky sap beetle is about 30 days with three to four generations per year. Some females lay more than 300 eggs and live as long as 147 days. The strawberry sap beetle primarily attacks strawberries. Sap beetles also disseminate organisms that cause rots in the fruits. Some sap beetles bore into the fruit, devour a portion, and lay eggs. Larval damage is usually only slight and often goes unnoticed.

Pest Control Measures

Sanitation

It is helpful to harvest sweet corn, tomatoes, melons, berries and other produce immediately as soon as they ripen. Remove any damaged, diseased, and overripe fruits and vegetables from the area at regular intervals. The collecting of apples, peaches, melons, tomatoes, and other decomposing fruits and vegetables and by burying deep in the soil or destroying is needed to eliminate beetle food sources.

Baits

Research has shown that picnic beetles prefer banana, whole wheat bread dough, and muskmelon. As a bait, muskmelon rinds or pineapple scraps, sprinkled with a pesticide, kills the strawberry sap beetle and other scavenger beetles attracted to the fruits and vegetables. Take extra precautions to keep treated baits away from humans, domestic animals and other non-target organisms. Bait trapping shows some promise in the reduction of beetle populations. Place traps several feet away from the picnic table or outside the garden. Discard trap contents frequently, every three or four days, and rebait traps with pineapple scraps and a bait consisting of stale beer, vinegar, molasses and water with yeast. Research has shown that carbaryl (Sevin) hardly kills sap beetles. Malathion gives better control and azinphosmethyl (Guthion) the best control. However, no pesticides are legally labeled for home owner use.

Sprays

Raspberries can be protected somewhat with repeat sprays of malathion as sap beetles begin to enter the fields. Treat three to five days before the first picking date. Use malathion 25 percent WP at the rate of four to five pounds per 100 gallons of water. Do not use malathion liquid as it can cause burning of the plant leaves. There is a one-day waiting period between application of malathion and harvest. Some sweet corn growers have reported limited success in killing sap beetles with carbaryl (Sevin). Read the label and follow directions and safety precautions as to which crops can be sprayed and the waiting period interval for harvest to avoid illegal chemical residues.

This publication contains pesticide recommendations that are subject to change at any time. These recommendations are provided only as a guide. It is always the pesticide applicator's responsibility, by law, to read and follow all current label directions for the specific pesticide being used. Due to constantly changing labels and product registration, some of the recommendations given in this writing may no longer be legal by the time you read them. If any information in these recommendations disagrees with the label, the recommendation must be disregarded. No endorsement is intended for products mentioned, nor is criticism meant for products not mentioned. The author, The Ohio State University and Ohio State University Extension assume no liability resulting from the use of these recommendations.

DISEASES

Strawberry plants can fall victim to a number of diseases. The leaves may be attacked by Powdery Mildew, Leaf spot (caused by the fungus Sphaerella fragariae), Leaf blight (caused by the fungus *Phomopsis obscurans*), and by a variety of Slime molds. The crown and roots may fall victim to Red Stele, Verticillium wilt, Black Root Rot, and Nematodes. The fruits are subject to damage from Gray Mold, Rhizopus Rot, and Leather Rot. The plants can also develop disease from temperature extremes during winter.

High yields of quality strawberries require vigorous growth and healthy plants. Growth can be affected by many different factors such as soil fertility, lack of moisture, weeds, insects, and diseases. Growers can control all of these factors, while certain factors, such as weather conditions, cannot be controlled. Growers can also control plant growth characteristics, productivities, and blossom hardiness by selecting different varieties. A sound strawberry production program should include the control of all pests and the use of good cultural practices. A total program replaces most of the "luck factor" and results in excellent crop production yearly.

Strawberries are attacked by many diseases that vary widely in their destructiveness and distribution. Preventative disease control measures begin with planting disease-free stock with genetic resistance and using recommended cultural practices. The production of disease-free plants and fruits may also require preventative applications of fungicides. When using fungicides to control diseases of foliage and fruit, follow label directions and the time limit given for "last application before harvest."

Foliage Diseases

Powdery Mildew. Leaf symptoms appear as white patches of mycelium on the lower leaf surface. The entire lower leaf surface may be covered, the leaf edges roll upward, and purple to reddish areas develop on the lower leaf surface. Losses result from infection of flowers and fruits, while leaf infection has not significantly reduced yields. A high degree of resistance is present in many cultivars.

Leaf spots. This complex of foliage problems includes: Leaf Scorch (Red Spot), Leaf Spot, Purple Leaf Spot, and several other leaf diseases. The symptoms are characterised by numerous irregular purple blotches, round lesions with red rusty borders and white centers, or spreading lesions that cover nearly the entire leaflet. Leaf diseases survive the winter on infected leaves as conidia or other overwintering fungal structures. The new disease cycle begins when rain or wind disperse fungal material to new leaves. Mid-age leaves are most susceptible and protective fungicides are required when susceptible cultivars are grown. Resistant cultivars of pathogen-free stock are desired.

Slime molds. Slime mold fungi may be found on strawberry plants and mulch during the warm, wet weather of spring or fall. This creamy white or tan colored, jellylike mass moves out of the soil and up onto leaves and grasses. It produces variously shaped and colored, crusty structures with powdery dry spores. Slime molds are found on fruit grown under plastic. These fungi are not parasitic to plants but can smother single leaves or fruits and be unsightly. Slime molds disappear when hot dry weather returns and do not require treatment.

Root and Crown Diseases

Red Stele. Diseased plants occur in patches and produce fewer runners and smaller berries. The tips of young roots rot first, after which the stele of infected roots becomes red. Lateral roots also die and decay, producing a "rattail" symptom. Red color may extend up to the plant crown. Plants may appear stunted or wilted, and leaf color may change to red, orange, or yellow. Avoid soil compaction and improve drainage. Plant resistant cultivars from inspected planting stock.

Verticillium Wilt. Damage from this disease is often most severe in the first year of growth. The oldest leaves show marginal and interveinal (between the veins) browning and death, while newer leaves are stunted but remain green and don't wilt. The symptom often appears in late spring with the onset of hot temperatures and dry periods, or with stress from high light intensity. Planting disease-free stock in soil without a history of Verticillium has been successful. However, this fungus has become widespread, so use of disease-resistant varieties and soil fumigation may be needed.

Black Root Rot. This is caused by a complex of root-attacking fungi, injurious environmental conditions (freezing or waterlogging of soil), nematodes, or any combination of these factors. Infected roots turn dark and lose their feeder roots. Black root rot causes poor plant vigor. This problem is associated with soils of high clay content, excessive irrigation, and soil compaction. Avoid these soils and provide adequate soil aeration for vigorous root growth.

Nematodes. Parasitic nematodes are small round worms that range in size from one-sixty-fourth to one-sixteenth of an inch long. Some species of nematodes cause serious damage to strawberries when they occur in high numbers. Root-knot nematodes cause characteristic knots or galls on roots.

Symptoms associated with root-knot and other nematodes include: stunting, yellowing of leaves, reduced berry yields, reduced production of runner plants, wilting, and general loss of vigor. Severe infestations of nematodes on strawberry plants have been found in a few instances in Minnesota.

Nematode root diseases are almost impossible to diagnose from symptoms alone. To make an accurate diagnosis, nematodes must be recovered and identified from diseased plants and soil. Such a diagnosis can be made only in a specially equipped laboratory by trained personnel.

The basic principle of nematode control in strawberries is planting only nematode-free strawberry plants in nematode-free soil. Eliminating nematodes from the soil is, at best, a rather expensive and difficult task. The basic method of field-scale nematode control is soil fumigation (application of chemicals). Crop rotation is generally more economical than soil fumigation if land is available.

Fruit Diseases

Gray Mold. This fruit disease develops as a cover of velvety gray mycelial and/or conidia on the fruit surface. The rot usually starts at a point of contact such as the soil or other infected fruits. The berry turns brown and remains firm with little leakage. Frequent rains produce maximum rot development, and fungicides from bloom to harvest are required. Mulching, debris removal, and narrower row widths help reduce this disease.

Rhizopus Rot. The berry is discolored, somewhat brown, softens quickly, and juices leak out when this disease occurs. If wet conditions continue, a dense white mycelium coats the berry. Wounds are necessary for penetration. Fungicides may be required during fruit ripening. Damage can also be reduced by quickly cooling berries after harvest.

Leather Rot. The berry may be affected from bloom to maturity. Young berries turn brown or dark brown while mature berries may be bleached, purple, or normal in color. The flesh remains firm and tastes bitter. Reduce fruit contact with soil and select fungicides for specific activity against this disease.

Noninfectious Disease

Winter injury. Some growers have lost many plants in new plantings and surviving plants were often weakened from winter injury. Low yields generally resulted. Injury is most likely to occur if plants are exposed to alternating warm and cold temperatures during winter. Mulching with hay, straw, leaves, or similar material reduces the chance of winter injury.

Fungicide Application

The production of the best possible crop from existing plantings may require fungicide treatment. Foliar and fruit diseases respond best to fungicide treatments, while root and crown problems, along with noninfectious problems, are best controlled by planting disease-free stock that is specific to the grower's area.Treatments should begin at bud break and, when required, may continue until harvest. Fungicides can be applied during bloom, which reduces blossom infection by fungi and results in fewer fruit blemishes. Repeat applications when needed at 7- to 10-day intervals up to the limitation given on the label. This will produce the best results. Fungicides must be used at label rates. Harvest applications generally are not necessary if several earlier applications were made.

PRODUCTION TRENDS

World strawberries production 2005-2007 in metric tons (only countries with production of more than 50000 tonnes in 2007 are mentioned):

Country	2005	2006	2007
USA	1,053,242	1,090,436	1,115,000
Russia	221,000	235,500	324,000
Spain	320,853	333,500	263,900
Turkey	200,000	211,127	239,076
Korea, South	201,995	205,307	200,000
Japan	196,200	190,600	193,000
Poland	184,627	193,666	168,200
Mexico	162,627	154,893	160,000
Germany	146,500	173,230	153,000
Egypt	100,000	105,000	104,000
Morocco	118,600	112,000	100,000
United Kingdom	68,600	65,900	66,000
Ukraine	46,200	47,800	63,000
Italy	146,769	131,305	57,670
France	57,637	57,221	57,500
Total world	3,789,701	3,908,978	3,822,989

USES

In addition to being consumed fresh, strawberries can be frozen, made into preserves, as well as dried and used in such things as cereal bars. Strawberries are a popular addition to dairy products, as in strawberry flavored ice cream, milkshakes, smoothies and yogurts. Strawberry pie is also popular.

Strawberry pigment extract can be used as a natural acid/base indicator due to the different color of the conjugate acid and conjugate base of the pigment.

Freezing

Strawberries are easy to freeze using a dry sugar or syrup pack. The dry-sugar pack is especially easy and gives the best flavor and color for sliced or crushed berries. For whole frozen berries a syrup pack is recommended because it produces a plump, well-shaped berry after thawing. For special sugar-free diets, strawberries can be frozen unsweetened, but they will not be as high in quality as sugar- or syrup-packed berries.

Twelve pounds or eight quarts of fresh strawberries will yield approximately 13 pints of frozen berries. No matter which type of pack you choose, follow these general directions for preparing and packaging strawberries for freezing:

- Use only firm, fully ripe berries.
- To avoid bruising and waterlogging the berries, wash only a few at a time in cold water.
- Drain on absorbent paper or in a colander or sieve.
- Remove the hulls with the tip of a fioating blade peeler.
- Chill the fruit in ice water to lower the temperature for fast freezing.

Syrup Pack

Make a syrup using 1-1/4 cups water to each cup of sugar. Dissolve the sugar in either cold or hot water. If hot water is used, be sure to chill the syrup before using. Use about 1/2 to 1/3 cup of sugar for each pint container. Place whole or sliced berries in containers and cover with cold syrup. Package and freeze.

Unsweetened Pack

Pack whole, sliced or crushed berries in containers and cover with water or berry juice. For better color retention, add ascorbic acid to the water, berry juice, or crushed berries. Cover crushed berries with their own juice.

Strawberry thrives best in temperate climate. It is a short day plant, which requires exposure to about 10 days of less than 8 hours sunshine for initiation of flowering. In winter, the plants do not make any growth and remain dormant. The exposure to low temperature during this period helps in breaking dormancy of the plant. In spring when the days become longer and the temperature rises. The plants resume growth and begin flowering. The varieties grown in milder subtropical climate do not require chilling and continue to make some growth during winter.

REFERENCES

Bostanian, N. J., M. Binns, J. Kovach, G. Racette and G Mailloux. 1999. "A predictive model for strawberry bud weevil adults in strawberry fields". *Environ. Entomol* 28: 388-406.

English-Loeb, G., M. P. Pritts, J. Kovach, R. Rieckenberg and M. J. Kelly. 1999. "Compensatory ability of strawberries for bud and flower removal: Implications for managing the strawberry bud weevil". *J. Econ. Entomol.* 92: 915-921.

Kovach, J., R. Rieckenberg, G. English-Loeb, M. Pritts. 1999. "Oviposition patterns of the strawberry bud weevil (Coleoptera: Curculionidae) at two spatial scales and implications for management." *J. Econ. Entomol.* 92: 1358-1363.

Stevenson Whitcomb Fletcher, *Strawberry Growing*, The Macmillan Co., New York, 1917.

Part II : Cultivation of Temperate Vegetables

Part II : Cultivation of Temperate Vegetables

10

Asparagus

Asparagus officinalis is a flowering plant species in the genus *Asparagus* from which the vegetable known as asparagus is obtained. It is native to most of Europe, northern Africa and western Asia. It is now also widely cultivated as a vegetable crop. Asparagus is a herbaceous perennial plant growing to 100-150 centimetres (39-59 in) tall, with stout larissa stems with much-branched feathery foliage. The "leaves" are in fact needle-like cladodes (modified stems) in the axils of scale leaves; they are 6-32 millimetres (0.24-1.3 in) long and 1 millimetre (0.039 in) broad, and clustered 4-15 together. The flowers are bell-shaped, greenish-white to yellowish, 4.5-6.5 millimetres (0.18-0.26 in) long, with six tepals partially fused together at the base; they are produced singly or in clusters of 2-3 in the junctions of the branchlets. It is usually dioecious, with male and female flowers on separate plants, but sometimes hermaphrodite flowers are found. The fruit is a small red berry 6-10 mm diameter.

Figure 1. Asparagus officinalis

Plants native to the western coasts of Europe (from northern Spain north to Ireland, Great Britain, and northwest Germany) are treated as *Asparagus officinalis* subsp. *prostratus*

(Dumort.) Corb., distinguished by its low-growing, often prostrate stems growing to only 30-70 centimetres (12-28 in) high, and shorter cladodes 2-18 millimetres (0.079-0.71 in) long. It is treated as a distinct species Asparagus prostratus Dumort by some authors.

Asparagus has been used from early times as a vegetable and medicine, owing to its delicate flavour and diuretic properties. There is a recipe for cooking asparagus in the oldest surviving book of recipes, Apicius's third century AD De re coquinaria, Book III. It was cultivated by the ancient Egyptians, Greeks and Romans, who ate it fresh when in season and dried the vegetable for use in winter. It lost its popularity in the Middle Ages but returned to favour in the seventeenth century.

COMMERCIAL PRODUCTION

Asparagus is an ancient vegetable, having been cultivated for over 2000 years. An ancient Roman author, Cato the Elder, in 200 B.C. wrote detailed directions for growing asparagus that are close to current recommendations.

There are a number of species, but only one, *Asparagus officinalis L.*, is cultivated for food. Asparagus is a favorite spring vegetable, both for its earliness and its distinctive taste. Not only is it low in calories, asparagus is an excellent source of vitamin A and contains significant amounts of calcium, phosphorus, riboflavin, and vitamin C.

Although commercial asparagus production in the United States is concentrated in California, Washington, and Michigan, many other areas have potential for producing asparagus. In South Dakota, the eastern end of the state and the Black Hills are two areas where asparagus can be grown most successfully.

The biggest problem in growing asparagus commercially is not the cultural practices but the marketing of the product. Partly because of competition from duty-free imports from Mexico and Peru, more of the U.S. production is now going into fresh market rather than processed. Although Americans are now consuming more asparagus than ever, the fresh market during May and June is nearly saturated, which has tended to keep prices low in recent years.

Nevertheless, a grower within a 15-minute drive of a city may be able to successfully market as a "pick-your-own," with one acre per 5,000 city dwellers. Asparagus also may be harvested and graded according to the specifications of the local supermarket and sold wholesale to them at, or retail through a farmers' market. The specific product marketing outlet, along with a realistic price, should be identified by a grower long before the first asparagus plant is put into the ground.

As of 2007, Peru is the world's leading asparagus exporter, followed by China and Mexico. The top asparagus importers were the United States (92,405 tonnes), followed by the European Union (external trade) (18,565 tonnes), and Japan (17,148 tonnes). The

United States' production for 2005 was on 218.5 square kilometres (54,000 acres) and yielded 90,200 tonnes, making it the world's third largest producer, after China (5,906,000 tonnes) and Peru (206,030 tonnes). U.S. production was concentrated in California, Michigan, and Washington. The crop is significant enough in California's Sacramento-San Joaquin River Delta region that the city of Stockton holds a festival every year to celebrate it, as does the city of Hart, Michigan, complete with a parade and asparagus queen. The Vale of Evesham in Worcestershire is heralded as the largest producer within Northern Europe, celebrating like Stockton, with a week long festival every year involving auctions of the best crop and locals dressing up as spears of asparagus as part of the British Aparagus Festival.

Asparagus was also corrupted in some places to "sparrow grass"; indeed, the Oxford English Dictionary quotes John Walker as having written in 1791 that "Sparrow-grass is so general that asparagus has an air of stiffness and pedantry". In Gloucestershire and Worcestershire it is also known simply as "grass". Another known colloquial variation of the term, most common in parts of Texas, is "aspar grass" or "asper grass". In the Midwest United States and Appalachia, "spar grass" is a common colloquialism. Asparagus is commonly known in fruit retail circles as "Sparrows Guts", etymologically distinct from the old term "sparrow grass", thus showing convergent language evolution.

It is known in French and Dutch as asperge, in Italian as *asparago*, in Portuguese as espargo hortense, in Spanish as *espárrago*, in German as Spargel, in Hungarian as spárga.

The Sanskrit name of Asparagus is Shatavari and it has been historically used in India as a part of Ayúrvedic medicines.In a South Indian language, Kannada, it is known as Ashadhi, Majjigegadde or Sipariberuballi.

CULTIVATION

Since asparagus often originates in maritime habitats, it thrives in soils that are too saline for normal weeds to grow in. Thus a little salt was traditionally used to suppress weeds in beds intended for asparagus; this has the disadvantage that the soil cannot be used for anything else. Some places are better for growing asparagus than others. The fertility of the soil is a large factor. "Crowns" are planted in winter, and the first shoots appear in spring; the first pickings or "thinnings" are known as sprue asparagus. Sprue have thin stems.

White asparagus, known as spargel, is cultivated by denying the plants light and increasing the amount of ultraviolet light the plants are exposed to while they are being grown. Less bitter than the green variety, it is very popular in the Netherlands, France, Belgium and Germany where 57,000 tonnes (61% of consumer demands) are produced annually.

Purple asparagus differs from its green and white counterparts, having high sugar and low fibre levels. Purple asparagus was originally developed in Italy and commercialised under the variety name *Violetto d'Albenga*. Since then, breeding work has continued in countries such as the United States and New Zealand.

In northwestern Europe, the season for asparagus production is short, traditionally beginning on April 23rd and ending on Midsummer Day.

Asparagus is a useful companion plant for tomatoes. The tomato plant repels the asparagus beetle, as do several other common companion plants of tomatoes, meanwhile asparagus may repel some harmful root nematodes that affect tomato plants.

Few vegetable dishes please the palate as much as fresh asparagus lightly steamed and with a dash of butter. Few vegetables are easier to grow, and no others are perennials that will return for 10-20 years once established. Rich in B vitamins, vitamin C, calcium, and iron, asparagus is often the first spring vegetable that you can harvest.

Asparagus thrives in full sun and prefers a light, well-drained soil. It will not tolerate competition; so prepare the planting area carefully and keep it mulched after planting. Asparagus is a heavy feeder that prefers a fairly neutral pH of 6.5-7.5. Generous applications of compost will provide the necessary nutrients as well as help to balance our acidic East Texas soils.

As a general rule, asparagus is established by planting one-year-old crowns. Choose your site carefully, preferably on the northern side of the garden to avoid shading out other plants. Loosen the soil 16" deep and amend it well with compost and rotted manure. Fall preparation for spring planting is best, but spring preparation will work unless the pH of your soil is very low. Plant the crowns about 6" deep and 1 1/2 ' apart, cover with soil, and mulch well. Keep the bed moist, but not wet. Raised beds are preferable to ensure that weeds and grass cannot encroach.

Don't plan to harvest any spears the first year. Allow the plants to grow lush foliage, which will be cut back in the fall after turning brown. Harvest lightly the second year, choosing two or three of the heavier spears per plant. Allow the wispy spears to grow into ferns once again. The third year, you can eat all you can cut, and the wait will be well worth it. Keep in mind that asparagus produces early, but still needs to be fed, watered and kept well mulched throughout the growing season. Treat it kindly and it will reward you with many years of delicious meals.

Keep the field free of weeds. The spears are less brittle in the afternoon and cultivation is best done at that time. Weeding problems can be simplified by using herbicides in the rows. Shield asparagus plants from spring prevailing winds to protect the spears from damage by blowing soil particles. Hot, dry winds can retard growth on the windward side of the spear causing production of crooked spears that often are unmarketable.

Leave the tops standing in the field in winter months; cut and burn (reduces insects and diseases) or disk them into the soil in the early spring before the shoots develop. In this area, the tops can help protect the roots from low temperatures by holding snow that prevents deep freezing and rapid scil temperature changes in the winter.

Planting

Asparagus is a perennial plant that can grow to a height of five feet or more and can produce spears for twelve to fifteen years. You should plant it an area where it will not shade other vegetables in your garden. It is grown for its edible spears, but the asparagus foliage is also a pleasing addition to a garden. This is a vegetable that requires patience, as the asparagus spears should not be harvested until the third year after being planted.

To prepare a bed for asparagus crowns, till an area about four feet wide and as long as you wish, at least 10 to 12 inches deep. Spread 3 to 4 inches of organic matter over the area and add manure at the rate of 1/2 pound per square foot. Add a complete fertiliser such as 10-20-10 at the rate of 1/2 pound per 100 square feet of bed area. Check the pH of your soil; and if is less than 6.0, add agricultural lime at the rate of 2-6 pounds per100 square feet of bed. Mix all of these ingredients thoroughly into the soil, and your bed is ready for planting, which is generally done during late winter or early spring.

After cold temperatures have abated, obtain asparagus crowns that are one to two years old, and plant them as soon as possible so they don't dry out. It is possible to keep them in a plastic bag in the refrigerator up to a week.

In your prepared bed, dig planting trenches not less than 10 inches deep and 10 inches wide, spaced about 2 1/2 to 3 feet apart. At the bottom of the trench, add two inches of a mixture of half compost and half well-rotted manure and about one teaspoon of phosphate fertiliser for each foot of row. Then add an inch of good soil to avoid placing the crowns directly onto the fertiliser. Slightly mound the soil down the center of the trench, and you are ready to plant your crowns.

Place the crowns over the mound in the trench 15 to 18 inches apart with the roots spread in an outward direction. Immediately cover the crowns with 2 inches of soil to prevent them from drying out. If you plant two rows, try to stagger the crowns so that they are not directly across from each other in the rows. After placing 2 inches of soil over the crowns, water well. In several weeks, the first young spears will appear. As they grow, you should fill in the trench until it is level with the garden surface, being careful not to cover the asparagus foliage. When the trench is completely filled, mulch around the plants with two inches of organic matter and keep weeds away from the plants.

Caring of Plants

To ensure strong root development and vigorous foliage growth during the first two years after planting an asparagus bed, the plants will require an inch of water per week during the growing season. In addition, prior to the emergence of spears, you should apply a nitrogen fertiliser, such as ammonium sulfate, at the rate of about 1/4 pound per 100 square feet. Manure may also be scattered over the surface of the bed at the rate of 50 pounds per 100 square feet.

At the end of the first growing season, to encourage production of spears, you should cut back the tops of the asparagus foliage to the ground. At this time, you may wish to dig up and remove the female plants, which produce smaller and fewer spears. These can be identified by the appearance of small red berries. If you dig out the female plants, you can replace them in the spring with healthy male plants.

During the first two growing seasons, keep the asparagus watered, fertilised, and weeded, and remove the tops in the winter, but do not harvest the spears. Two years after planting one-year-old crowns, you can begin harvesting spears when they are 6 inches tall. With a sharp knife, cut diagonally through the spear just below ground level in order to prevent the emergence of a clump of woody stumps above ground. To encourage strengthening of the root storage system, the first harvest season should last only about four weeks. Thereafter, the harvest season can be extended to up to eight weeks. A large proportion of the spears will emerge during the first half of the harvest season, and production will taper off as the season progresses. Harvesting should cease when the spears are less than 3/8 inch in diameter.

Climatic Requirements

To be productive, the asparagus plant needs a long rest period during winter and an average summer growing temperature of 66-75 F. The climate of this area meets these requirements. The crowns are hardy and are seldom injured by cold. Winter injury usually is restricted to fields where the tops have been removed in the fall.

Field Selection

When selecting a field for asparagus production, choose one that has not been in asparagus for the last eight years. Good soil drainage is essential, and the highest yields are obtained on deep sandy loams, since asparagus roots may extend six feet deep. The field should be fairly level to avoid soil erosion. Avoid shaded areas or low-lying areas subject to late spring frosts that could delay harvest and reduce yield. The soil should be free from stones, which can cause crooked spears. The field should also be free of perennial weeds, which are difficult to control as there are few herbicides available for establishing asparagus. Heavy soils that form a crust after rain or irrigation will damage the emerging spears. Do not plant in muck soils since freezing in these soils heaves and damages the crown.

Varieties

Crowns of the Washington cultivars (e.g., Mary or Waltham Washington) have been most readily available, especially to home growers, although newer hybrids are becoming more common. Take advantage of the new male hybrids from New Jersey, such as Jersey Giant or Jersey King. These hybrids have been developed to have greater rust disease resistance and higher yield, along with good cold hardiness, and have done well in trials in North Dakota and Minnesota. A novel variety that has shown promise in Iowa trials is "Purple Passion," which has purplish-colored large-diameter spears.

Selecting Crowns

Crowns may be purchased from a reputable nursery or produced from seed in a separate bed for later transplanting. Choose soil that is well drained, deep, loose, and light. Generally, one acre of seedlings will produce enough crowns for ten acres of asparagus. There are 900-1200 seeds per ounce. One ounce of seed can produce 550 high-quality crowns. You can figure out how many crowns you need and plant the seed accordingly.

To hasten germination, soak the seed for 48 to 96 hours in warm water before planting; change the water frequently to avoid fungal growth. It is also advisable to use a seed fungicide. Plant seed one-half inch deep, 2-3 inches apart in rows 24-30 inches apart. Optimal temperatures for germination are 60-85 degrees F.

Cultivate during the growing season to control weeds. When seedlings are one year old, they may be transplanted to their permanent location. Dig the crowns in the spring before the buds begin to grow. Old plant tops can be mown if they interfere with the digging. A modified potato digger or moldboard plow can be used to lift crowns from larger plantings. One-year-old crowns are more vigorous and have greater productivity as compared to 2 to 3 year old crowns that suffer more root damage during digging, which may reduce quality. Good-quality crowns weigh 150 pounds per 1,000.

Transplant Production

An alternative to planting one-year-old crowns is to use transplants, which can be grown in a greenhouse. To do this, seed can be treated as above but then seeded into peat pots filled with a good potting media (not field soil). Plant two seeds per pot, and thin to one after emergence. Transplants can be started either in the early spring for setting out after danger of frost has passed or in late summer for transplanting in mid-Sept. Transplant when seedlings are 10 to 12 weeks old into the permanent bed prepared as for crowns: Pots should be covered with at least one inch of soil at the time of planting in the furrows and the furrows filled in gradually over the season as with crowns. Apply a transplant starter fertiliser solution at the time of planting.

Discard any small or badly injured crowns. Sort crowns according to size. Plant crowns of the same size in a row to avoid competition for food. If needed, crowns can be stored at 40 F degrees and 85-90% humidity until planting. Do not allow to freeze. To avoid decay, the crowns should not have any damp or moist areas on them during storage. Before planting, dip the crown in a recommended fungicide solution. Before planting the crowns, prepare the land so the soil is free of stumps, rocks and other coarse rubbish that might cause crooked spears. Any existing weeds should be controlled; spraying with a glyphosate herbicide two to three weeks before planting is recommended.

Open the planting furrow to a depth of 8-10 inches. Position the crowns, upright with the roots spreading, in the furrow about 8 inches below ground level and then cover with about 2 inches of soil. Close spacing may be practiced in the home garden, but when planting half an acre or more, it's usual to space rows 3 to 4 ½ ft with plants 12-18 inches apart in the row.

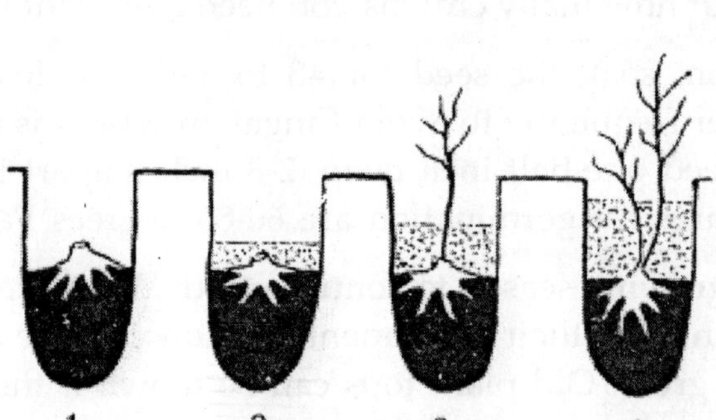

As the crowns grow, gradually work the soil toward the plants during cultivation until the furrow is entirely filled. Be careful not to bury the ferns. Each cultivation should add about 2 inches of soil over the crowns. About three cultivations should cover the crowns and completely fill the furrow. Shallowplanted crowns often are injured by spring frosts or by disking and produce smaller spears. Deeply planted crowns produce larger spears that emerge later in the spring.

Fertiliser

Since asparagus occupies the land for 15-20 years, good soil preparation before planting is important. If possible, apply 5-10 tons of good farmyard manure per acre or plow down a green manure crop in the fall. Otherwise apply 600-1,000 pounds of 8-32-16 per acre (20 lbs per 1,000 square feet or ½ lb per 10 ft of row) before planting. A soil test is the best way to determine fertiliser needs. Also make a test for boron, particularly in lighter soil. The soil should contain 0.5 ppm of boron. A boron application of two pounds per acre should correct deficiencies.

Apply 50 lbs of nitrogen per acre each year after planting. Also apply 50 lbs of phosphorus and potassium every other year to maintain a good yield. Broadcast fertiliser and then either water in or work into the soil by shallow cultivation. Regardless of the general fertiliser practice, it is desirable to split the nitrogen application, applying half (25 pounds) in the spring before growth starts and half at the end of the harvest season (0.5 cup of ammonium nitrate [35-0-0] per 10 ft of row in spring and after harvest). The second application gives the plant more vegetative growth to accumulate more carbohydrates which gives a higher yield the following year and improves plant vigor to survive the winter.

Crooked Spears

Crooked spears generally are the result of dirt clods in the field, hot dry wind in early spring, or injury due to insects, frost, disking, or heavy soils. Avoid all these problems to produce straight, marketable spears.

Irrigation

Irrigation is not essential for established asparagus in this area, but it will increase productivity (spear size and numbers) and longevity of the planting. It will also increase vigor and decrease risk of infection by Fusarium, a fungal crown rot. Apply water at least for the first 2 years, if possible.

Male vs. Female Plants

Asparagus is dioecious. Non-hybrid varieties will have about 50% male and 50% female plants, but all-male hybrids are available. Female plants are less productive and have a shorter life because the plant's energy is allocated to seed production in the fall. Male plants produce a larger number of spears, starting earlier in spring, but they often are thin. Female plants produce fewer spears, but they are larger and are produced later in the season. The Jersey male hybrid varieties do not produce seeds.

Diseases and Problems

Since fusarium root rot and crown rot eventually invade every planting, cultural methods that maintain the vigor of the plants will help prolong productivity.

Fusarium root rot: The fungus lives in the soil and enters the plant through the roots. Plant asparagus in a field that did not have asparagus for the last eight years, and dip the crowns in recommended fungicide before planting. Keep the planting healthy by practicing good cultural methods.

Rust: Cut and burn the tops in the spring, and use resistant varieties, such as the Jersey hybrids. Fungicides may be used in commercial plantings.

Feathering: Bracts of the spears are partly spread as a result of high temperatures.

Harvest

Delay harvest until two years after planting, and then limit harvest in the third year to 2-4 weeks, depending on the vigor of the fern the previous season. Established fields may provide a yield of 2,000 to 4,000 pounds per acre (3/4 to 1 pound per foot of row in a home garden), depending on climate and cultural practices.

Snap spears by hand or cut 2 inches below ground level, at about a 45 degree angle, when they are 6-10 inches tall. Morning harvest is preferred because the spears then contain a maximum amount of water and stay fresh longer. The frequency of harvesting depends on temperature: spear growth is faster with higher temperatures. Studies in Michigan showed 22-23 pickings per season (about six weeks) optimal for both yield and spear size over time, compared to fewer (15) or more (28) pickings. This will vary somewhat depending on local climate.

After harvest, store asparagus at cool temperatures, between 32-36 F to maintain quality.

Lignin fibers make asparagus tough and undesirable. The tip of the spear has less lignin and the amount increases toward the bottom. The following practices can reduce lignin in the spears: Snap asparagus, rather than cutting at ground level, thus leaving most of the fiber portion (white portion) of the spear in the field. During cold weather, harvest shorter spears. The longer it takes a spear to grow, the more fibers it develops. Eliminate small diameter (thin) spears; they have more fiber on a weight basis. Cool spears as quickly as possible after harvest in ice cold water. Store asparagus at less than 36 F degrees, but do not allow it to freeze. Avoid water stress by wrapping asparagus in a wet material or in perforated plastic wrap, or store upright in a bowl with the stems in water. Do not store asparagus for more than 7-10 days.

USES

Culinary

Three types of asparagus on a shop display, with white asparagus at the back and green asparagus in the middle. The plant at the front is Ornithogalum pyrenaicum, is commonly called wild asparagus, and sometimes "Bath Asparagus". Only the young shoots of asparagus are eaten.

Asparagus is low in calories, contains no fat or cholesterol, and is very low in sodium. It is a good source of folic acid, potassium, dietary fiber, and rutin. The amino acid asparagine gets its name from asparagus, the asparagus plant being rich in this compound.

The shoots are prepared and served in a number of ways around the world. In Asian-style cooking, asparagus is often stir-fried. Cantonese restaurants in the United States often serve asparagus stir-fried with chicken, shrimp, or beef, also wrapped in bacon. Asparagus may also be quickly grilled over charcoal or hardwood embers. It is also used as an ingredient in some stews and soups. In the French style, it is often boiled or steamed and served with hollandaise sauce, melted butter or olive oil, Parmesan cheese or mayonnaise. It may even be used in a dessert. The best asparagus tends to be early growth and is often simply steamed and served along with melted butter. Tall, narrow asparagus cooking pots allow the shoots to be steamed gently, their tips staying out of the water.

Asparagus can also be pickled and stored for several years. Some brands may label them as "marinated" which means the same thing.

The bottom portion of asparagus often contains sand and dirt and as such thorough cleaning is generally advised in cooking asparagus.

Green asparagus is eaten worldwide, though the availability of imports throughout the year has made it less of a delicacy than it once was. However, in the UK, due to the short growing season and demand for local produce, asparagus commands a premium and the "asparagus season is a highlight of the foodie calendar." In continental northern Europe, there is also a strong seasonal following for local white asparagus, nicknamed "white gold".

Medicinal

Ingestion of asparagus may bring on an attack of gout in certain individuals due to the high level of purines.Following the consumption of asparagus, the urine of some people contains various sulfur-containing compounds, which produce a smell identified by some people. The prevalence of smell production and detection is debated.

This phenomenon has been observed since the 18th century, and is caused by asparagusic acid and its derivatives, which are present in asparagus but not other vegetables.

The effect of eating asparagus on the eater's urine has long been observed:

> "asparagus... affects the urine with a foetid smell and therefore have been suspected by some physicians as not friendly to the kidneys; when they are older, and begin to ramify, they lose this quality; but then they are not so agreeable".

Marcel Proust claimed that asparagus "...transforms my chamber-pot into a flask of perfume."

Certain compounds in asparagus are metabolised giving urine a distinctive smell due to various sulfur-containing degradation products, including various thiols, thioesters, and ammonia. The volatile organic compounds responsible for the smell are identified as:

- methanethiol,
- dimethyl sulfide,
- dimethyl disulfide,
- bis(methylthio)methane,
- dimethyl sulfoxide, and
- dimethyl sulfone.

Subjectively, the first two are the most pungent, while the last two (sulfur-oxidised) give a sweet aroma. A mixture of these compounds form a "reconstituted asparagus urine" odor. This was first investigated in 1891 by Marceli Nencki, who attributed the smell to methanethiol. These compounds originate in the asparagus as asparagusic acid and its derivatives, as these are the only sulfur-containing compounds unique to asparagus. As these are more present in young asparagus, this accords with the observation that the smell is more pronounced after eating young asparagus.The biological mechanism for the production of these compounds is less clear.

REFERENCES

Grubben, G.J.H. & Denton, O.A. (2004) *Plant Resources of Tropical Africa 2. Vegetables.* PROTA Foundation, Wageningen; Backhuys, Leiden; CTA, Wageningen.

Blamey, M. & Grey-Wilson, C. (1989). *Flora of Britain and Northern Europe.*

Vaughan, J.G.; Geissler, C.A. (1997). *The New Oxford Book of Food Plants.* Oxford University Press.

11

Broccoli

Broccoli is from the brassica family, which include cauliflower, cabbage, Brussels sprouts and the Chinese brassicas like bok choy. It's a cool season vegetable and forms florets rather than a single large head like a cauliflower which can also withstand warmer temperatures. If you're wondering about your climate for growing broccoli, it will tolerate frosts of 20°F (6.5°C) and likes daytime temperatures around 50-70F° F (10-20°C) and certainly no higher than 80°F (26°C), or it will bolt to seed.

Figure 1. Broccoli

Depending on your climate, sow seeds early to mid summer, which will take 4-6 weeks before ready for transplanting, or plant seedlings out no later than late summer. Broccoli needs to develop a good size with strong leaf growth before the cool weather sets in. If you get your broccoli growing early enough without suffering excessive heat, it will be ready for use in late autumn and winter.

VARIETIES

There are three commonly grown types of broccoli. The most familiar is sometimes called calabrese in Great Britain and simply "broccoli" in North America. It has large (10 to 20 cm) green heads and thick stalks and is named after Calabria in Italy. It is a cool season annual crop.

Sprouting broccoli has a larger number of heads with many thin stalks. It is planted in May to be harvested during the winter or early the following year in temperate climates. The heirloom variety "calabrese" available in North America is of this type.

Romanesco broccoli has a distinctive fractal appearance of its heads, and is yellow-green in colour. It is technically in the Botrytis (cauliflower) cultivar group.

Purple cauliflower is a type of broccoli sold in southern Italy, Spain, and the United Kingdom. It has a head shaped like cauliflower, but consisting of tiny flower buds. It sometimes, but not always, has a purple cast to the tips of the flower buds.

CULTIVATION

Broccoli is a cool-weather crop that does poorly in hot summer weather. Broccoli grows best when exposed to an average daily temperature between 65 and 75 degrees Fahrenheit (18 to 23 degrees Celsius). The majority (99%) of the United States broccoli crop is grown in California and Arizona. Other cultivar groups of Brassica oleracea include cabbage (Capitata Group), cauliflower (Botrytis Group), kale and collard greens (Acephala Group), kohlrabi (Gongylodes Group), and Brussels sprouts (Gemmifera Group). Chinese broccoli (Alboglabra Group) is also a cultivar group of *Brassica oleracea.*

Broccoli is high in vitamins C, K, and A, as well as dietary fiber; it also contains multiple nutrients with potent anti-cancer properties, such as diindolylmethane and selenium. A single serving provides more than 30 mg of Vitamin C and a half-cup provides 52 mg of Vitamin C. The 3,3'-Diindolylmethane only found in broccoli is a potent modulator of the innate immune response system with anti-viral, anti-bacterial and anti-cancer activity. Broccoli also contains the compound glucoraphanin, which can be processed into an anticancer compound sulforaphane, though the benefits of broccoli are greatly reduced if the vegetable is boiled more than ten minutes. A high intake of broccoli has been found to reduce the risk of aggressive prostate cancer. Broccoli leaf is also edible and contains far more betacarotene than the florets.

Broccoli is usually boiled or steamed, but may be eaten raw and has become popular as a raw vegetable in hors-d'oeuvre trays. Although boiling has been shown to reduce the levels of suspected anticancer compounds in broccoli, other preparation methods such as steaming, microwaving, and stir-frying have been shown not to reduce the presence of these compounds.

FAO reports that in 2008, China and India were the top producers of cauliflower and broccoli. About half of all cauliflower is raised in China and one fourth in India. In Europe, production has been in Spain since about 1990. In North America, production is primarily in California. The seasonal average f.o.b. shipping-poit price for cauliflower in 2004 was $33.00 per 100 pounds ($0.73/kg) according to the National Agricultural Statistics Service, USDA.

Tips on Growing broccoli

You will not need many plants as each plant will produce continuously over time, but do consider successive plantings as you don't want it all to mature at the same time. Usually 8-10 plants are enough for the average family over a season. Allow 3-4 weeks between plantings to space the harvest over time.

Sow your broccoli seeds in punnets and prick out the less hardy plants as they sprout. Once they are 7-10cm high (3-4 inches) they can be planted into your garden bed. The bed should have good drainage and be well prepared with lots of organic compost beforehand.

Seedlings should be placed at least 45cm (18") apart. Ensure the winter sun reaches them for at least 5 hours a day.

Watch for caterpillars and aphids as they tend to like to invite themselves to dinner on your broccoli leaves. For solutions to common pest problems see pest control. But here's a handy hint: Before broccoli and cabbages form heads, sprinkle the leaves with cayenne pepper powder. This should deter the little blighters.

Broccoli is a leafy green plant so growing it fast with regular top ups of organic nitrogen fertiliser is a must. Fertilise every 2-3 weeks.

Certain varieties of broccoli are better for warmer areas than others, so check locally for the ideal plants for your garden.

If you have gone overboard with growing broccoli and end up with more than you can use; cut it, blanch and freeze it for use later. It tends to loose its freshness in the refrigerator after a few days.

Growing Broccoli is easy and rewarding and because it has proven health benefits learning how to grow Broccoli is well worth the effort.

The green frozen heads you find in shops are really called calabrese and not strictly Broccoli.

The name Broccoli is generally reserved for sprouting varieties which when cut will sprout again....but boy are they delicious.

Growing Broccoli? - then sow in the spring, transplant in the summer and harvest during the following spring, a welcome gap filler for that time of year.

Soil

An important part of growing Broccoli is understanding your soils make-up. Ideally Purple Sprouting Broccoli likes a firm, rich, heavy soil (but not clay) as these tend to be higher in nutrients.

Broccoli plants are going to be in the ground for approximately 38-42 weeks so the ground should not be waterlogged in winter.

Pick an area that is sunny or partially shaded and fork in some well rotted compost during the the autumn/fall. This need not be done every year just make sure your soil is kept in good heart as part of your ongoing care.

About 2 weeks before sowing apply a general fertiliser, tread down - not too heavily though - then lightly rake over the surface.

Your soils pH is a measure of its alkalinity or acidity and different plants thrive at different pH levels. In reality though most plants cope well across a range of readings.

You can measure your soils pH with a testing kit bought from a garden centre or plant nursery. Growing Broccoli in a soil with a pH range of 6.2 to 7.2 (7 is best) will be just fine.

Adding lime will increase your pH reading... making your soil more alkaline. Attempting to make your soil more acidic - a lower pH reading - is more difficult.

Sowing and Planting

As you can see these steps on how to grow Broccoli are easy and do-able for your home garden.

Different Broccoli varieties may have differing sowing times so all you need to do is read the seed packet...carefully - and you won't go wrong. But as a general 'rule of thumb' both white and Purple Sprouting Broccoli plants are sown in April/May time and transplanted June/July. Let's go deeper - don't worry we are not going too deep... just up to your ankles.

Making a Seed Bed

In early spring rake over and level the area you have chosen as a seed bed - frosts will have done a lot of the work for you by then... hopefully. If you only require a few plants then this need not be too big an area.

Cover the bed with sifted mature compost - the bed may be small enough to purchase a bag or two from your garden centre, especially if you're just getting growing Broccoli, you might not have the organic matter to compost.

Tread over this to make it firm... but only when the soil is dry enough i.e. not sticking to your boots. Then give it a light raking over.

Mark out the rows using string stretched between two stakes for a straight line and make shallow drills along the length of the string with a trowel 13mm(1/2") deep and rows 15cm(6") apart.

Sow your Broccoli seed thinly into the drills and cover them over with soil, lightly firming with the head of the rake. Mark both ends of the row, labelling one end with the variety and date of sowing. Keep the seed bed moist, always use a fine spray.

When the growing Broccoli plants are about 2.5cm(1") tall, thin the seedlings to about 7.5cm(3") apart in the rows. This is to stop overcrowding thus causing the seedlings to become weak and spindly.

The seedlings are ready to plant out when they are about 10cm(4") to 15cm(6") tall. Water the bed the previous day before removing the Broccoli seedlings to their permanent position.

When your learning how to grow Broccoli there is no need to remember which varieties to grow. Part of the enjoyment of growing Broccoli and indeed, all vegetables, is to get hold of some of those amazing seed catalogues (they're sent out free by seed companies) and get to know what is available - then make your choice... especially around a nice cosy fire. The anticipation really energises you for the coming spring.

To harvest Broccoli plants a few weeks earlier place a cloche over your seeded rows, or sow the seeds in 7cm(3") pots of seed compost on a window ledge or greenhouse - this will bring them on a lot quicker.

Transplant them into their permanent beds when they are 10-15cm(4-6") tall. An important fact of knowing how to grow broccoli successfully is that... just like many of us, they like a firm bed - so transplant them firmly, 2.5cm(1") deeper than the growing Broccoli were when in the seed bed, leaving 45cm(18") between plants. Water the hole you have created for the vegetable prior to planting.

Sowing into Trays

As ever, when learning how to grow Broccoli or any vegetable, there are other ways to achieve the same outcome.

So if you have a greenhouse, cold frame or cloches - growing Broccoli from seed can be easier and more convenient if you don't have a large garden to accomodate a seed bed.

Sow Broccoli seed in a tray filled with seed compost bought from your local gardening centre. Water thoroughly and place them inside one of the above (cloche etc...).

When the first two leaves have formed prick them out into 7cm(3") pots filled with potting compost. Plant them into these pots a little deeper - to just below the two leaves - water in well.

Leave plants to grow on until tall enough for planting out by following the growing instructions above.

If your not just growing Broccoli and you want a more detailed description of how to sow seed into trays then follow this link - Starting Seeds - The page will open into a new browser window - the instructions are about halfway down the page.

Taking Care of your Plants

After-care is very much part of learning how to grow Broccoli - but fortunately it's easy:

- Carefully hoe around the plants to keep weeds down.
- Protect from those pesky birds - they love the young seedlings. Put canes in the ground with upturned plant pots on top. Drape netting over and secure with bricks on the ground.
- Water regularly especially in dry conditions.
- Apply a mulch to conserve moisture - grass clippings on top of wet news papers, well rotted compost or manure - to mention but a few.
- From time to time feed the growing Broccoli with a liquid fertiliser.
- As the plants get taller you may need to stake the plants.
- Re-firm the Broccoli plants if the have become loosend by wind or frost.

Pest Control

Growing Broccoli whilst minimising pest problems will mean following some sort of crop rotation plan. This simply means not growing vegetables of the same family in the same piece of earth year after year...it encourages pest build up in the soil. For Broccoli this also means - Brussels Sprouts, Cabbage, Kohlrabi, Turnip, Cauliflower.

Cabbage Root Fly can be a problem so use protective discs at the base of the growing Broccoli as they lay their eggs in the soil at this point and when they hatch the little maggots burrow down and feed on the roots.

These discs can be made from old bits of hession backed carpet, underlay or roofing felt - use your imagination, or purchase them ready made from a garden nursery or DIY store.

Also cover with gardening fleece when the plants are young. Fleece is a great friend to the gardener, it allows rain and light through but insulates against frost and deters pests.

Keep a watchfull eye for the appearance of Caterpillars as left unattended they will devastate your growing Broccoli plants. They feed on the underside of the leaves and the best answere is to just pick them off and... squish 'em.

They also lay their eggs on the underside of the leaves too so look for yellow clusters aannndddd... squish 'em. If you you have'nt the time to keep squishing - or the stomach, then use 'Just Caterpillar'. It uses nematodes to provide an environmentally safe and efficient treatment which can be watered straight onto edible plants.

Use 'Just Caterpillar' once they are seen on the plant, preferably whilst they are still small, and the nematodes will quickly seek out the caterpillar and kill it. Get it from your garden centre. Spraying with Derris will also have an impact.

It's a blood thirsty game learning how to grow Broccoli - the carnage is terrible. ;0)

Club Root is a fungus and all brassicas (thats the family umberalla for Broccoli along with Brussels Sprouts, Cabbage, Kohlrabi, Turnip, Cauliflower) can be affected.

The roots of your young growing Broccoli become stubby and swollen. Leaves become yellow and wilt causing severe stunting of growth. This causes swelling and reduces the flow of water and nutrients to the plant. Spores are produced and can survive in the soil for up to five years.

Prevent club root by practicing crop rotation and take extreme care if you buy in seedlings from outside as this is often the way this fungus is introduced.

Reducing the acidity of the soil by adding lime will help.

Aphids can be a serious pest by weakening your growing Broccoli plants and introducing viruses. Try planting Marigolds or Tagetes plants among the crop. They attract beneficial insects like hover flies and ladybirds which feed on the Aphids and will help reduce the infestation.

Knowing how nature operates helps us understand how to grow Broccoli and other vegetables in a way that helps your environment positively. But spraying is the only way to make a real impact on badly infested plants.

Spray with Derris - an organic compound recommended by organic growers.

Flea Beatles are tiny beetles that can make sieves out of your brassica leaves. When you touch the Broccoli leaves they ping off just like regular fleas, only these won't bite you! To deter them:

Use horticultural fleece placed over your broccoli, as soon as you transplant them outside; a floating row cover.

Lightly hoe over the soil regularly to destroy eggs and larvae and expose them to predators.

Give your plants a mid-day shower with the hose (not in full sun though), as they're most active then and they don't like wet conditions.

Try 'Companion Planting', to attract the beneficial insects.

Vast changes have taken place in the UK and Europe concerning the availability of pesticides during recent years to the extent that most have been removed from retailers shelves. Visit your garden centre or nursery and seek their advice as what to use.

When your learning how to grow Broccoli, it is important to be able to recognise the pests from the goodies - you don't want to kill the goodies if you can help it. So observation is the way to to become knowledgeable... it all adds to the enjoyment.

Harvesting

This final step on how to grow Broccoli is always the best step. It means you are about to be rewarded for all your efforts - and the long wait... but it's all been worth it.

The time to cut your first 'spear' or flower shoot is when they are well formed but the tiny flower buds have not yet opened. Letting them flower will render the spear tasteless.

To produce the side shoots which will keep on appearing you first need to cut the main central spear of your growing Broccoli. As these side shoots grow (10-15cm(4-6") long), you will want to pick them regularly. Picking can go on for 6 weeks but if they flower, shoot production will stop very quickly.

REFERENCES

Buck, P. A.. "Origin and Taxonomy of Broccoli" (in English). Department of Food Technology, University of California. http://www.springerlink.com/content/ert85x3082740212/fulltext.pdf. Retrieved on 2009-05-14.

"Broccoli Cultivation Factsheet, Clemson University extension (2003)". http://hgic.clemson.edu/factsheets/HGIC1301.htm. Retrieved on 2008-05-13.

Dixon, G.R. (2007). *Vegetable brassicas and related crucifers*. Wallingford: CABI.

Warwick Medical School, University of Warwick (2007-05-15). "Research Says Boiling Broccoli Ruins Its Anti Cancer Properties".

12

Capsicum

Capsicum (or pepper in the US, Canada and United Kingdom) is a genus of plants from the nightshade family (Solanaceae) native to the Americas, where it was cultivated for thousands of years by the people of the tropical Americas, and is now cultivated worldwide. Some of the members of Capsicum are used as spices, vegetables, and medicines. The fruit of Capsicum plants have a variety of names depending on place and type. They are commonly called chilli pepper, red or green pepper, or sweet pepper in Britain, and typically just capsicum in Australian and Indian English. The large mild form is called bell pepper in the US and Canada. They are called paprika in some other countries (although paprika can also refer to the powdered spice made from various capsicum fruit).

The original Mexican term, chilli (now chile in Mexico) came from the Nahuatl word chilli or xilli, referring to a larger Capsicum variety cultivated at least since 3000 BC, as evidenced by remains found in pottery from Puebla and Oaxaca.

The fruit of most species of Capsicum contains capsaicin (methyl vanillyl nonenamide), a lipophilic chemical that can produce a strong burning sensation in the mouth of the unaccustomed eater. Most mammals find this unpleasant; however, birds are unaffected. The secretion of capsaicin is an adaptation to protect the fruit from consumption by mammals while the bright colors attract birds that will spread the seeds.

The amount of capsaicin in Capsicums is highly variable and dependent on genetics, giving almost all types of Capsicums varied amounts of perceived heat. The only Capsicum without capsaicin is the bell pepper, a cultivar of *Capsicum annuum*, which has a zero rating on the Scoville scale.

Chili peppers are of great importance in Native American medicine, and capsaicin is used in modern medicine-mainly in topical medications-as a circulatory stimulant and pain reliever. In more recent times, an aerosol extract of capsaicin, usually known as capsicum or pepper spray, has become widely used by police forces as a non-lethal means of incapacitating a person and in more widely dispersed for riot control, or by individuals for personal defence.

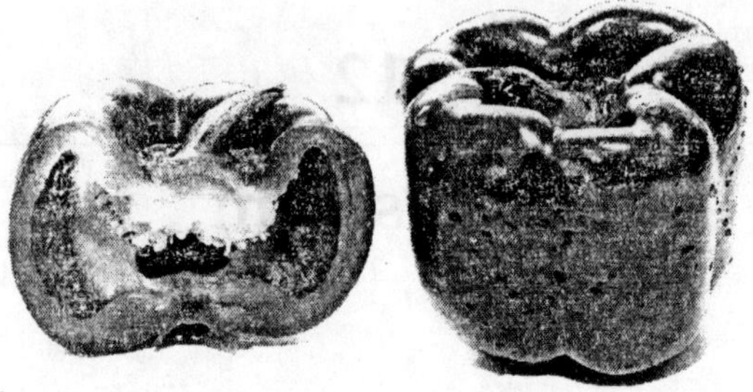

Figure 1.Capsicum

Although black pepper and Sichuan pepper cause similar burning sensations, they are caused by different substances-piperine and hydroxy-alpha sanshool, respectively.

SPECIES AND VARIETIES

There are many modern varieties (cultivars) of greenhouse capsicum available. Main characteristics are the fruit shape and fruit colour (red, yellow, orange or other, while green capsicums are usually unripe-red capsicums); resistances (most modern varieties are tolerant for one or more virus types); production, fruit quality and vigour. Some varieties are very vigorous and plants can become as tall as 3.5 m in one year. Regularly new varieties (cultivars) are developed and it is worthwhile comparing a few plants of a new variety against the standard one.

Capsicum consists of approximately 20-27 species, five of which are domesticated: C. annuum, *C. baccatum*, *C. chinense*, *C. frutescens*, and *C. pubescens*. Phylogenetic relationships between species were investigated using biogeographical,[9] morphological, chemosystematic, hybridisation, and genetic data. Fruits of Capsicum can vary tremendously in color, shape, and size both between and within species, which has led to confusion over the relationships between taxa. Chemosystematic studies helped distinguish the difference between varieties and species. For example, *C. baccatum* var. baccatum had the same flavonoids as *C. baccatum* var. pendulum, which led researchers to believe that the two groups belonged to the same species.

Many varieties of the same species can be used in many different ways; for example, *C. annuum* includes the "bell pepper" variety, which is sold in both its immature green state and its red, yellow or orange ripe state. This same species has other varieties as well, such as the Anaheim chiles often used for stuffing, the dried Ancho chile used to make chili powder, the mild-to-hot Jalapeño, and the smoked, ripe Jalapeño, known as a Chipotle.

Most of the capsaicin in a pungent (hot) pepper is concentrated in blisters on the epidermis of the interior ribs (septa) that divide the chambers of the fruit to which the seeds are attached. A study on capsaicin production in fruits of C. chinense showed that capsaicinoids are produced only in the epidermal cells of the interlocular septa of pungent fruits, that blister formation only occurs as a result of capsaicinoid accumulation, and that pungency and blister formation are controlled by a single locus, Pun1, for which there exist at least two recessive alleles that result in non-pungency of *C. chinense* fruits.

The amount of capsaicin in hot peppers varies significantly between varieties, and is measured in Scoville heat units (SHU).

GROWING GREENHOUSE CAPSICUMS

Capsicum, or sweet pepper, is a valuable crop with excellent prospects both for the domestic and export market. It has very good potential as an alternative for greenhouse tomatoes. However, growing capsicum is harder and more laborious than growing tomatoes.

Capsicum plants are slow-growing, and it may take over five months from sowing to first harvest. They need a high temperature to develop at a reasonable rate. Fruit set occurs in periods and fruit are harvested in flushes.

They can be grown perfectly in soil, growing media or NFT, but they are sensitive to salt and require good water. Capsicum is prone to disorders like blossom-end rot and cracking, as well as to some pests and diseases.

The price per kg is usually high, but the production rather low: overseas top productions are in the order of 25 kg/m² per annum for red capsicum and 30 kg/m² for green capsicum, compared to over 60 kg/m² for tomatoes.

Very hygienic working practices are needed to avoid spread of virus. It is important that ali capsicums produced are of a very high quality standard. This requires a high level of knowledge, commitment and investment of the grower.

Sowing

Always use fresh, high-quality virus-free seeds. The seed count is 100-140 per gram, and one gram of seed will generally raise about 80-100 plants. For soil-grown crops sometimes broadcast sowing is applied with 1-2 gram per m². Normally sowing is done in trays with 10 gram per m² for scattered sowing, or 40 gram per m² for space-sowing.

Sowing individual seeds in pots requires much more space, but the advantage is that no pricking out is needed. For rockwool, seedlings are sown either in propagation cubes or in rockwool flocks and later transplanted to cubes. Seed trays are filled with a high-quality soilless mix, peat, vermiculite, perlite or rockwool flocks. A complete fertiliser must be given, or in rockwool a normal nutrient solution is given but with a CF of 20 (EC of 2). The seeds must be covered with a thin layer (6 mm) of fine-grade medium, and watered with a fine spray. The soil or the trays must be covered with glass or plastic to avoid drying out.

At 28°C emergence will occur about a week earlier than at 21°C. The optimal germination temperature is 24-25°C. The young seedlings require good light conditions. Directly after germination the covering material is removed and the temperature lowered to 23°C, although 18 to 25°C is possible. The seedlings can be pricked out when the cotyledons are fully expanded.

Propagation

Capsicum growth is very much dependent on temperature. Particularly the root temperature must be sufficiently high (20-22°C). So for propagation under cold conditions the use of benches or bottom heat is beneficial. Capsicum transplants require a minimum day temperature of 21 - 23°C, and two degrees higher during sunshine, and a night temperature of 20°C.

Humidity must be high, but overhead spraying is not advised as it increases the risk of Botrytis. CO2 enrichment (to 700 ppm) will promote growth of the seedlings from the stage where true leaves are expanded. Good light conditions are essential. Artificial lighting supports plant propagation, but whether this is economical depends on natural light levels, prices etc.

In growing media, nutrients should be available as soon as the seedlings start growing. In peat, liquid-feet should start four to six weeks after pricking out. The young seedlings can be placed on a small area, but once leaves begin to touch they must get more space (18 plants/m^2). The plants are often supported by two small sticks without an elastic band.

Planting and Spacing

Plants are ready for planting when the first flower bud is visible. They have then reached a fresh weight of 35 - 45 gram. This will be five weeks after sowing in summer, or up to eight weeks under unfavourable conditions. In good planting conditions (warm peat) it is possible to plant relatively larger plants. If the conditions in the greenhouse are not optimal (low root temperature) it is better to plant smaller plants. The main root of the capsicum plant should be able to grow out freely without being obstructed by the bottom of a plastic pot.

The plant density in the greenhouse depends on the training system to be applied later (single-row, V-system, etc.). However, it is stem density rather than plant density that matters. Optimal stem density is around seven or eight stems per m^2 in summer (may be higher in a new glasshouse), and around six stem per m^2 in unfavourable light conditions. It is most common to have two main stems per plant, for which 3 - 4 plants/ m^2 should be planted.

The optimal temperature after planting is 22/21°C (day/night) on the first days, and 23/22°C later. Air humidity is best around 80 % or 4 g/m^3 deficit. The root zone temperature is very important and should be around 22°C. Lower root temperature increases the risk of Pythium, while too high root temperature will cause too much plant growth and insufficient flower production.

Good soil structure and moisture content are important, and high salt or high nutrient levels are unfavourable. The optimal CF is 30, or up to 40 when transpiration is low e.g. in dull weather. The young plants require a good calcium and nitrate supply. The roots will establish within a week, and preparations for plant training can start.

Plant Structure

Capsicum plants initially develop one single stem. (Fig 2)After 9 - 13 leaves, a terminal flower develops and the plant branches into two. Sometimes three or four shoots develop naturally rather than two. Also two flowers may develop rather than one. The flower in the first branching is the crown bud. Each shoot will branch further after one or two leaves, while producing flower buds at the point of division, and the plant grows further in this way. However, if no pruning is applied, the later side-branches will not develop, thus naturally avoiding a too bushy plant shape.

The most common way of growing capsicum is with two main shoots, as this is how most plants develop by themselves. If three or four main shoots are required, the third and fourth must normally be taken from the second branching. The advantage of three main shoots is that costs for plant material is lower, while production is about equal for two or three stems per plant, at the same stem density. The difficulty is to get three equal shoots, and it is very important that the sideshoots have equal vigour. The four stems training system requires a vigorous variety.

Training and Pruning

We describe the training for plants with two stems. After the first division, two shoots develop which will form the two stems. These shoots must be trained up strings to the crop wire. Either the stems are loosely (!) wound around the strings, or better still, the stems are clipped to the strings using rings.

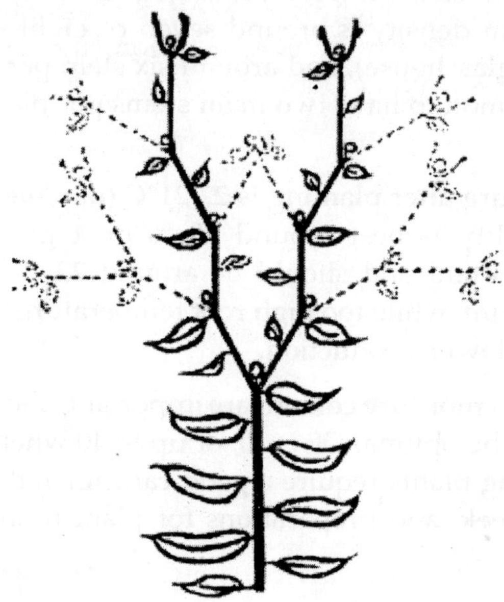

Figure 2. Pruning a plant with two stems

The plant will continue producing a terminal flower and two side shoots at every internode (Figure 2). Of every two new side shoots, one is retained to continue the stem, and the other is removed or just pruned, leaving one or a few leaves. Three leaves are left on the side shoot when the plant needs extra leaf area, either for photosynthesis or for shading the fruit. In older plants, the pruned side shoots will carry fruit.

Training and pruning should be done every three weeks, or even every two weeks in periods of fast growth. In summer it is better not to prune leaves in the upper 15 cm of the plant in order to provide shading to the fruit. At the end when the last fruit are set the plants can be topped to avoid stem breaking and to improve fruit size.

Flowering

A young capsicum plant starts flowering about two to six weeks after planting, when it has 7 to 13 leaves. The first flower, or crown flower, either aborts or must be removed. Later flowers may develop into fruit. The temperature during early flower development affects the size and the shape of the later capsicum fruit. The lower the temperature during flower development, the shorter the fruit (length compared to the diameter). A flower grown under very low temperature (below 10°C at night) would produce a small flattened fruit. A lower temperature during flowering results in less four-loculed fruit, and more three-loculed fruit, or even two-loculed fruit, which is not desired. Also, flowers developed at a night temperature below 18°C usually produce a fruit with a 'tail' (elongated, pointing blossom-end). In conclusion, the optimal temperature during flowering is 20 - 21°C on average both day and night.

Pollination

Capsicum can be self-pollinating, but there is a high degree of cross-pollination because honey bees, thrips and other insects transfer pollen from blossom to blossom. Pollination is not markedly improved by using an 'electric bee' nor by spraying plant hormones, but pollination is clearly better when honey bees or bumble bees fly in the greenhouse. Bees increase the number of seeds in capsicum fruit, and such fruit attract more assimilates, and therefore grow faster or bigger. Bees make that less fruit are deformed, and that fruit set and first harvest are better and a bit earlier.

Plant Condition for Fruit Set

In a young crop, fruit set should only be allowed when all conditions are good. Until then, all flowers and young fruit should be removed at an early stage. The main condition is that sufficient leaf area must have developed. Plants with three or four axils above the branching are big enough.

Secondly, light conditions must be good so that plenty of assimilates (sugars) can be produced at all times. Larger plants require a higher minimum light level, as they need more energy for their own maintenance. Thirdly, the flowers must be strong and in optimal condition, and best if they are bending downwards. Reducing the (night) temperature will improve the flower condition so that fruit set can take place after that.

Inducing Fruit Set

Fruit set will occur by itself when the light conditions are good. However, when fruit set is difficult there are several methods to overcome this. The most practical way is by dropping the night temperature to 18°C. The day temperature can stay at 20-22°C, but if fruit set needs further help, the day temperature can also be dropped to as low as 18 oC. It is good if the root temperature can be dropped to 18°C as it will give more flowers. Very low night temperature (10°C) ensures a high percentage of fruit set, but induces fruit with a few seeds only, which will abort or remain small. About 18°C night temperature (and root temperature) is optimal for fruit set.

Other ways to help fruit set are: increasing the CO_2 concentration (to 700-1000 ppm), and also increasing the CF (to 40). Day temperature and light are not so important for pollination itself, but high light and low day temperature are beneficial for setting and first growth of the new fruit. Fruit set takes about a week, and is normally completed in five to eight weeks after planting.

For setting of the second flush of capsicum fruit, dropping the temperature is not advised, as it will slow down the ripening of the first flush. Generally, the later fruit will set by themselves after harvest of the older fruit. Harvesting of the first flush must be done with some consideration as it significantly influences plant growth and later fruit set.

In an old crop, fruit set can be difficult especially when the light levels go down. In an old plant stand, it is important to allow for a long fruit growth period (6 - 9 weeks for green fruit, 9 - 12 weeks for red) and to get the last flush set as soon as possible. This can best be done by harvesting the plants empty, even harvesting green, and then setting night temperature very low (even as low as 14°C) until fruit set is completed. Raising the CO_2 level will help too. Fruit ripening can be sped up after the harvestable-green stage by higher temperature (24°C average).

Controlling Fruit Load

Each plant can raise a limited amount of fruit, either limited in number or in weight. Too much fruit load on a plant may cause decaying of roots, and as a consequence Pythium may attack. The number of fruit a plant can raise depends on light, temperature and plant size. Young plants can support only 12 fruit per m^2 under poor conditions, and over 24 fruit per m^2 under good light conditions (this is four to eight fruit per plant, or two to four per stem). An older crop can have over 40 fruits per m^2 at a time. Mature plants control fruit load themselves by fruit set and fruit thinning. In young plants, however, fruit load must be controlled by the grower, in order to achieve a reasonable fruit weight (e.g. minimal 150 gram per fruit).

The best method to correct the number of fruit is by removing excess. This improves fruit quality (as inferior fruit are taken out) and fruit size (as remaining fruit experience less competition). The earlier the fruit thinning is done, the more the other plant parts (fruit, shoots and roots) will benefit. However, if too few fruit are left on the plant, these are more at risk for fruit cracking.

Secondly, fruit load can be corrected by raising the temperature just after fruit set. This reduces the amount of assimilates available, which will cause abortion of some new fruit. However, this method does not have the positive effects on fruit quality. Thirdly, plant load is affected by harvesting.

Fruit Growth

About 5 - 9 weeks after fruit set, fruit will reach their full size and get firm, while still green (the 'harvestable-green' stage). Further ripening on the plant until fully-coloured takes another 2 - 4 weeks. The total period from blossom to full-colour ripe fruit is 7 - 12 weeks. The huge variation in fruit growth duration is due to many factors: season (light), temperature, fruit load, condition of the plant, plant management, cultivar, etc.

Fruit growth can be hastened by high temperature (around 24°C). Generally, one degree higher average temperature reduces the fruit growth period with two to three days, but high temperature also reduces the average fruit weight. High temperature is not recommended under poor light conditions. Too high temperature (above 28°C) stimulates blossom-end rot, and too low temperature (below 17°C) at early fruit growth may cause a 'tail' on the fruit.

Harvesting

Capsicum fruit can be harvested when they are harvestable-green. Green fruit are incapable of ripening after removal from the plant. When harvested a week after the harvestable-green stage the fruit will turn gradually into colour, which is not desired. The right stage for red fruit is when they have reached full colour and are still firm. Overripe fruit are soft, and have a low keeping quality.

Capsicum fruit must be harvested with a very sharp, small knife, to get a smooth stem-end appearance, and so to minimise damage to other fruit. Also, a small wound on the plant reduces the risk of Botrytis infection. If plants are suspected of having a virus infection, the harvest knife should be dipped in low-fat milk with 3.5 percent protein to capture the virus and avoid virus spreading to other plants.

For delivering good quality fruit, it is best to start harvesting early in the morning and to finish before the hottest hours of the day. It is most important that harvesting, handling and packaging are done with the greatest care. Capsicum fruit are very prone to handling damage, which usually becomes visible after storage.

Fruit will mature in flushes, certainly ... the beginning of the production. In peak periods frequent harvesting is needed: coloured fruit once or twice a week, and green fruit once per fortnight. After a harvest peak it is still good to make a quick harvest round e.g. once a fortnight to pick the odd late or early fruit.

Effect of Harvest on the Plant

Harvesting is 'unloading' the plant, and allowing energy to go into new growth. New shoots will develop and produce a new flush of flowers and young fruit. Harvesting green fruit will result in an earlier unloading, and thus earlier setting of new fruit. Green harvesting can be recommended for the first flush when the plants are still small. If only some fruit are harvested green, the other fruit are more at risk of cracking, so for that reason it is better to harvest all first fruit green at once.

On the other hand, when all fruits from one stem are harvested at the same time, the sudden change in fruit load may cause a too vigorous growth of that stem. It will produce flowers that are too coarse, which will develop in low quality fruit. So the last ripe fruit should be harvested after some new fruit have set on that stem. The effect of harvesting on shoot growth can be used to achieve equal length of the stems on one plant.

Temperature Control

Young plants generally require a high average temperature to stimulate the appearance of new leaves and development of leaf area. A typical minimum temperature for young capsicum plants would be around 22°C (21- 23) (average for day and night). This can be achieved with for instance 25/19°C (day/night), or with 22/22°C. The latter regime will give more compact plants. If the young plants are planted in summer under high light conditions, the average temperature can be set one or two degrees higher. Also during flowering the night temperature and the average temperature must be relatively high (minimum 20 - 21°C) for a good flower quality.

For good pollination and for getting fruit with many seeds a lower night temperature (18°C) is optimal. Day temperature can remain at 20-22°C, or can also drop. Directly after fruit set a moderate average temperature (20°C) is set for growth of young fruit, or a higher temperature (22°C or more) can be chosen to stimulate abortion of excessive fruit.

During early fruit growth the mean temperature (24 hour average) can be chosen between 20 and 23°C depending on the conditions. An average below 19°C may give rise to 'buttons' and 'tails'. A higher average temperature (23°C) hastens fruit growth, but reduces the ultimate fruit size. To achieve 20°C on average, often the day temperature is set around 23°C, and the temperature in the first half of the night is set low, e.g. 16°C, because a low night temperature gives bigger fruit and better keeping quality. The maximum temperature should ideally not come beyond 28°C.

The ideal method for heating would be by heating pipes, where the lower pipes heat the roots and are also used as rails for the trolleys. The higher pipes can be at the height of the fruit. In cold winter conditions it can be economic to use a thermal screen.

PLANT HEALTH OF GREENHOUSE CAPSICUMS

Plant health problems affecting capsicum crops can be divided into two groups: 1) pests and diseases that are caused by living organisms, and 2) disorders brought about by chemical or physical causes. Pests include insects and mites. Diseases are caused by pathogens including fungi, viruses, bacteria and nematodes. Disorders cover a variety of problems, like physiological disorders (e.g. sun scald, cracking, blossom-end rot), nutrient deficiency, nutrient toxicity, salt injury, cold injury, damage from pesticides (including herbicides) and damage from gases (ethylene, ozone). This is a brief overview of the main pest and disease problems in New Zealand capsicums.

Pests

Thrips: Small, long, thin, brown/black insects, with similar shaped larvae which are white/ yellow. Thrips feed on flowers and on the lower side of the leaves, causing silver or grey-white spots with black faecal dots. Another symptom is small 'warts' on the lower leaf side, caused by egg deposition. High numbers of thrips will prevent flowers setting and can affect plant growth and reduce production.

Two thrips pests are found on New Zealand capsicums: onion thrips and western flower thrips. Both species can transmit tomato spotted wilt virus which can mark fruit and kill plants. Western flower thrips is resistant to most pesticides and for effective control it is important to know which thrips you have. Both thrips can be controlled by a predatory mite, *Amblyseius cucumeris*.

Two-spotted mite: These are tiny (0.5 mm) web spinning mites, coloured yellow brown with two dark marks on the sides. They feed on the lower side of the leaves, and cause very yellow feeding marks. In severe infestations whole leaves turn yellow and webbing will cover the plant. Successful biological control using the predatory mite *Phytoseiulus persimilis* depends on finding infestations of the pest early and releasing the predator immediately.

Broad mite and strawberry (or cyclamen) mite are so tiny that they are difficult to see, but they cause distinct virus-like symptoms (distortion of leaves) and can prevent new growth. If pesticides are used, the plant must be thoroughly wetted. *Amblyseius cucumeris* may control small infestations of these mites.

Below is an example of broad mite damage on capsicum fruit. The marking on the surface of the fruit was caused by mites feeding when the fruit was young.

Aphids: Various aphids are fond of capsicum plants, especially the young leaves in the top of the plants. Symptoms of aphids include sticky honey dew, white empty skins on the leaves, and feeding injury in the form of curling leaf margins, yellow spots, leaf distortion. Severe infection may cause leaf drop and as a result low production and sun scald on fruit. Aphids stain the fruit with their excrement and may transmit viruses.

Caterpillars: Green loopers are the commonest caterpillar damaging capsicum plants. The first signs of caterpillars are holes in the leaves. Large caterpillars may damage fruit. Green looper and tomato fruit worm are controlled effectively using a biological spray containing *Bacillus thuringiensis* toxins. This is a selective and environmental-friendly spray. The most damaging and difficult to control caterpillar is the poroporo fruit borer, a native species that fortunately only occasionally gets in to crops. The white or pink caterpillars can be found in the fruit.

Whitefly: This is not a severe problem in capsicum. Adults and larvae feed on plant sap and excrete a sticky honey dew staining leaves and fruit. They can be controlled by a parasitic wasp, *Encarsia formosa.*

Fungal Diseases - Affecting Roots

Three common root diseases are associated with soil that remains too wet for long periods. This is usually the result of poor drainage. So improving drainage is the primary solution. Some of the root diseases can develop in hydroponics. To kill inoculum of all three disease, the soil or medium must be sterilised before the next crop is planted.

Pythium and Rhizoctonia: are two fungi causing damping-off of seedlings. They arise when plants are weak or under stress (too cold, too wet), also in older crops. Generally Pythium is more active at lower temperature, and Rhizoctonia at higher temperatures. Pythium thrives under wet conditions, also in hydroponics. Symptoms of Pythium are brown roots: while the inside root is rot, the root skin is still intact and can be easily removed. Rhizoctonia typically has brown lesions at soil level.

Phytophthora: infects the main root and the lower stem part. Later the above-ground stem part turns black and the plant wilts and decays. Phytophthora causes soft rot (watery rot). It spreads through soil and water. Infected plants must be removed carefully and good sanitation practices are required.

Fungal Diseases - Above-Ground

Botrytis (grey mould): attacks many crops. It thrives under humid conditions, and needs very high humidity or free water for germination. Infection may occur on leaves, flowers, fruit and stems, mainly in stem wounds. Symptoms are brown lesions, expanding rapidly, and later symptoms are a grey mass of fine fungal tissue on dead plant material. The pathogen is resistant to many fungicides and for their effective use, resistance management strategies must be practised. Botrytis infection can be minimised by keeping plants dry and maintaining low humidity (also at night) by heating and venting at the same time (especially early morning!).

Downy mildew: shows as a white fungal tissue on the lower side of leaves, and later also on the upper side. Early symptoms are light green spots on the upper side of the leaves (similar to spider mite symptoms). These spots grow and later cover the whole leaf until it falls off.

Sclerotinia: causes shoots to turn brown, often without others signs of fungal infection. When there is enough water, a dense white fungal tissue will develop. It can also affect fruit. Infected material must be removed. When the infection is serious, the soil or medium must be sterilised.

Viral Diseases

Capsicum plants can be infected by several viruses, including tomato (or tobacco) mosaic virus (mechanically transmitted) and tomato spotted wilt virus (transmitted by thrips). Viruses cannot be seen but are detected by plant symptoms which include mottling, distortion or discolouring (flecking, ring patters, mosaic) of leaves and fruit, and stunted or bushy plant shape. It may reduce the production and fruit quality. A virus disease in the crop can not be cured. It is therefore important to prevent virus infection, or when it is detected, to reduce spreading.

It is important to identify the cause of a problem so that informed actions can be taken, because some virus-like symptoms are not caused by a virus but by mites. Also, viruses are transmitted in different ways (mechanical, by aphids or by thrips). So identification and hygiene are the key.

Mechanically transmitted viruses can be spread by knives, hands, clothing, boxes, etc. Viruses can stay alive for over a year in living and dead plant material e.g. in seed, old plant or roots parts in the system, or in soil or water. When there was virus in the old crop, the greenhouse needs to be thoroughly cleaned and sterilised before planting the new crop. Remove all plant material, even small parts, from the greenhouse and dispose or completely isolate them. Sterilise the medium, greenhouse, nutrition recycling system, boxes, sorting machine, all tools, shoes, etc.; thoroughly clean the whole work area including the cafeteria; hot-wash all clothes. Removal of all weeds from inside the greenhouse and the surrounds will reduce the risk from insect borne viruses and from insect infestation.

The new crop must be planted with all possible precautions. Always start from high-quality seed, as inferior seed is a major source of virus. Propagate the plants elsewhere in a greenhouse with insect screening, and have them arrive when your property is perfectly clean. After planting, be alert for the virus symptoms and have these identified. Any infected plant plus the neighbouring plants should be taken out, put in a plastic bag on the spot, and disposed carefully.

If a virus has established, some practices will help to prevent spreading and minimise the damage. Control insects if these are a vector. Wash hands in soapy water and dip knives in trisodium phosphate, Virkon, or milk containing at least 3.5% proteins. The milk must be renewed daily. For all work in the crop, always go in one direction. Have large shoe disinfection trays at the entrance of your property and at every greenhouse entrance. Make enough time for the big clean-up after ending the crop.

Diseases Caused by Bacteria

Several bacterial diseases may affect capsicums. They can cause soft rot, wilting, spots, deformation, leaf drop, and thus devastate the production and quality. Transmission may be by seed, insects or mechanically. It is very difficult to control bacterial diseases in a crop. Prevention is the only viable approach and seedlings and young plants should be treated with full precautions to prevent bacterial diseases.

Diseases Caused by Nematodes

Nematodes are tiny worms normally living in the soil. Some of them may feed on plant roots. One species can cause small knots or galls on the roots of capsicum and other plants. Above ground symptoms may include stunting, yellowing, wilting, lack of vigour. Nematodes can be killed by good fumigation or steaming between crops, but this must be done after the roots are decayed.

REFERENCES

Heiser Jr, C.B.; Pickersgill, B. (1969). "Names for the Cultivated Capsicum Species (Solanaceae)". *Taxon* 18 (3): 277–283.

Eshbaugh, W.H. (1970). "A Biosystematic and Evolutionary Study of Capsicum baccatum (Solanaceae)". *Brittonia* 22 (1): 31–43.

Capsicums: Innovative Uses of an Ancient Crop History, Botany, Breeding, and Pungency. Purdue University, Indiana, U.S.A.

13

Lettuce

The terms "specialty" and "gourmet" mean different things to different people when applied to lettuce, and in some markets any high-quality lettuce that isn't an iceberg is a specialty crop. Specialty lettuces and "mesclun" began receiving a lot of interest from both producers and consumers in the late 1980s. Consumer interest in salads generally, but especially in unusual or "new" salads such as mesclun, increased rapidly, and prices soared as demand outstripped supply.Such high prices and rapidly growing markets always attract a lot of attention from growers seeking greater returns per acre, and the resulting increase in production eventually exerts downward pressure on prices. Specialty lettuces are no exception to this phenomenon.

Organic agriculture has now come under federal regulation. All growers wishing to label and market their produce as organic must become certified; only those small farmers who market less than $5,000 of organic products annually are exempt from this requirement. While federal regulation of organics is new, the essential nature of organic farming remains the same. Organic production still entails the growing of crops without synthetic pesticides or standard commercial fertilisers. Organic growers utilise a wide range of cultural practices and alternative inputs to manage crops in a manner believed to be safer for the environment and better for the consumer.

CULTIVATION

While a mesclun mix may contain any mixture of lettuces, greens, and herbs, young leaf lettuces are normally the primary component. Leaf lettuces are generally tastier and less watery than head lettuce, while the variety of leaf shapes (smooth to crinkled edges) and colors (pale green to red) makes them attractive in leafy salads. Chefs like the texture of the leaves, which hold dressing well. Growers like these lettuces because they are generally the earliest maturing, are relatively trouble-free, and may resprout if cut young as a baby lettuce. Many "old fashioned" looseleaf lettuce cultivars are consequently making a comeback as specialty lettuces. Of particular note are the "oakleaf" types, with their attractive leaf shape and texture.

Figure 1. Lettuces

Apart from the standard iceberg-type lettuces, there are a number of lettuce "families" that may or may not be considered gourmet or specialty crops, depending on how wide a definition is used. The closer to the top of the following list a family is, the more "mainstream" it is, meaning that its price will usually be lower; the market for that type of lettuce, however, will be deeper than the market for types lower on the list. There are both green and red members of most of these families.

- Batavian lettuces are somewhat more open than icebergs, but form a distinct head nevertheless; they tend to be more flavorful than iceberg.
- Romaine types are upright and deep-colored, with a robust flavor and somewhat coarse texture.
- Bibb varieties are low-growing, tender, and sweet, forming a small, loose head as they approach maturity. Bibbs are sometimes called "Boston lettuce" or "Butter lettuce."
- Smooth-leaf lettuces are open-growing and tender, with generally pale leaves and delicate flavor when young.
- Crinkle-leaf sorts are similar to smooth-leaf lettuces except that their leaves are more "frilly" and their flavor is often slightly more pronounced.
- Unusual types have particular characteristics, such as a pronounced oakleaf shape, differentiating them from the other families.

Several non-lettuce species are also regarded as specialty greens and may be grown either alone (for very limited markets) or with an assortment of small lettuces. The major types of non-lettuces include:

- Arugula, a plant of the mustard family; when eaten the initial flavor is somewhat peanutty, followed by spicy-hot.
- Mâche, while well known in Europe, is uncommon in North America outside Québec. With a delicate nutty flavor, the mildest of any greens, mâche will get lost in almost any blend and is best sold alone.
- Sorrel is also well known in Europe, but is almost totally unfamiliar to consumers this side of the Atlantic. It is a tart cool-weather plant, most commonly used in potato-based soups.
- Chicory embodies a range of species, including radicchio, witloof (Belgian endive), and leafy chicories. All have a bitter flavor and very particular cultural requirements.
- Oriental greens include a number of pungent species, most of them in the mustard and chrysanthemum families. Some examples are mizuna, tatsoi, and red mustard.
- Unusual types encompass several species (dandelion, fennel, and so on) best known in their traditional ethnic markets. They are particularly useful for adding an accent to otherwise ordinary mixes.

Many of the cultural techniques for lettuce and salad-mix production are similar, regardless of whether the crop is grown in a greenhouse or in the field. Contact Cooperative Extension or seed suppliers for varietal recommendations specific to your growing region. Take note that organically grown seed is required for organic production. Conventionally-grown untreated seed is permitted only if its organic equivalent is not available. Fungicide-treated seed is completely prohibited.

Soil Preparation

Lettuce prefers a sandy-loam soil high in organic matter. Lettuce is sensitive to soil acidity; lime should be added, if necessary, to adjust the pH to 6.5-7.0. Lettuce is also sensitive to excess salts, especially at germination. If the previous crop was heavily fertilised, leaching the soil through heavy irrigation might be advisable. Field-grown lettuce, especially, requires good drainage to avoid certain fungal diseases such as bottom rot. Certified organic production requires that fertility be managed without the use of conventional fertilisers.

When you plant your lettuce, you'll need to make sure to grow it in well-draining soil with plenty of organic compost mixed in. Sandy to loamy soils work well for lettuce. Keep your soil pH at around 6.5 for best results.

It's a good idea to take a soil test before growing lettuce. You can use an electronic soil tester to verify that your soil has the right pH, nutrients and soil moisture levels. This way you'll know if your soil needs some fixing, which is especially important if you're accustomed to using fertilisers in your garden. Many times we use fertilisers when they are not really necessary. Testing your soil will give you a better idea of what your plants really need.

If you add well-decomposed, high quality compost to your soil, your lettuce shouldn't need additional fertiliser. If you choose, try out an organic fertiliser or a mycorrhisal root builder. These natural products will boost your plants natural abilities to extract nutrients from the soil. According to a study by Texas A&M , organic fertilisers actually work better than chemical fertilisers, and are better for the environment.

Another excellent thing you can do for your lettuce is to place a layer of organic mulch over your garden after you've planted and your lettuce starts to develop. Use a 3-4 inch layer of bark chips, shredded wood, straw, etc. The mulch will break down over time, adding nutrients to the soil. Mulch will also help your soil retain moisture and will keep weeds from growing.

Planting Your Lettuce

Lettuce won't take up much space in your vegetable garden like tomatoes and cucumbers, so it's a good crop for smaller gardens. Make your rows at least 10 inches wide, with at least 20 inches between the rows. A 24 inch spacing is recommended for some varieties, so check the seed package for recommended spacing.

As your lettuce grows, you should monitor their growth and thin out your plants periodically. Butterleaf and romaine need about eight inches between each plant, whereas crisp lettuce needs about 12-14 inches.

You can plant most varieties of lettuce directly in the ground from seed. Romaine and butter lettuce grow well from both seeds or from transplants. However, crisphead lettuce is best when grown from transplants. Transplants are a popular pick for head lettuce as head lettuce planted in the spring by seed has a high rate of failure. Head lettuce is risky to plant with seeds as it is hard to time the harvest correctly, making it susceptible to bolting. Thus, it's worth buying transplants so that it will mature in time.

When planting lettuce seeds, place them fairly shallow in the soil. Cover your seeds with 3/8 - 1/2 inch of soil. After planting, give your seeds a good drink. If you wish, you can add an organic fungicide to the soil to help prevent problems with damping off and other fungal diseases.

In some parts of the U.S., you can plant your seeds or transplants in the garden as early as January. In colder areas, you'll need to wait until March or so. Young, tender lettuce seedlings are sensitive to freezing, but can typically handle a light freeze.

Plan on growing a fall crop of lettuce around three months before the first serious fall freeze. If you're going to plant seeds in the summer for a fall crop of lettuce, you can plant seeds directly in the garden without much risk. However, just make sure that you properly water your plants, especially if the weather is warm and dry when you plant the seeds. Check with your extension agent for special planting times in your area.

Lettuce generally requires a lot of water. Keep you plants well irrigated unless you're getting a decent amount of rain.

Types of Lettuce

Most gardeners and farmers group lettuce into four different types.

Romaines (Cos)

Romaine (cos) is a tall lettuce with rigid leaves that are tightly packed together. The leaves are dark green, with the inner leaves having less color. Some consider romaine the most flavorful lettuce variety.

Crispheads

Crisphead lettuce is a very common variety of lettuce to find in your grocery store. Iceberg lettuce is a crisphead lettuce. The leaves are thin, light green, and densely packed together. It is best to grow this lettuce in cooler weather or start with transplants because of problems with bolting.

Butterheads

Butterhead lettuce is smaller than other lettuce varieties. The leaves are not as tightly packed in as with crisphead lettuce, and the leaves are more tender than crisphead leaves. This lettuce has a smooth, light, and some would say "buttery" flavor.

Leaf lettuce

Leaf lettuces are the most varied of the lettuce varieties. Colors can vary from red to dark green. The leaves can be curly or ruffled. This is a favorite of home gardeners as it matures quickly once planted.

Propagation

Lettuce production frequently entails transplanting. There are a variety of methods for raising transplants. Most growers prefer plug trays for seedling production. Since the potting mix does not have to be screened thoroughly, as blocking mixes do, commercially available organic potting- soil products can be used. This saves the screening and mixing time necessary when making your own soil-blocking mix. Since many commercially-available potting mixes contain wetting agents and chemical fertilisers, be certain the product is organic before purchasing or consider making your own with approved ingredients.

If a large percentage of the crop is sold as mesclun, "baby-lettuce," or salad mix, many growers will seed specialty lettuce directly and skip the transplanting stage—after all, the market crop is often no bigger than a robust transplant. Transplanting is a more realistic option in situations where the target market desires full-sized greens. Lettuce seed germinates better if left uncovered. Germination temperature should be between 15 and 18°C (60-65°F), with relative humidity around 70%. Some small-scale producers have expressed a preference for soil blocks—ejection-molded cubes of specially prepared potting mix. If you choose to adopt this transplant production system, soil blockers are available through Johnny's Selected Seeds and other sources.

Crop Scheduling

The primary challenge of specialty lettuce growing is to ensure a steady supply through the marketing season; careful planning and timing of planting dates is critical, and difficult to do well without considerable experience growing lettuces and similar greens. If the plan is to maintain a continuous supply of lettuce throughout the growing season, it is necessary to follow a different planting schedule than one might expect. Lettuces, like most plants, grow and mature at varying rates during different seasons. In cooler temperatures, short days, and the low light levels of autumn and winter, it is especially challenging to maintain a reasonably steady supply. In summer, for lettuce in the field, it is the length and severity of the hot dry period of midsummer that becomes a problem. The maturity time (planting to harvest) will double or triple for plantings to be harvested from September through February, so fall planting dates should be adjusted accordingly. It may be desirable to harvest every week, but this does not necessarily mean that crops should be planted at seven-day intervals.

The following schedule, recommended by a greenhouse lettuce producer in Britain, provides an example of seeding intervals required to time weekly harvesting from early November through April. The seemingly conflicting intervals probably reflect adjustments for varying day length, which has an effect on the growth rate of winter greens.

September 1-10	sow every 3 ½ days
September 10-18	sow every 2 days
September 18-October 10	sow every 3 ½ days
October 10-November 15	sow every 7 days
November 15-December 15	sow every 10 days

Adapt this schedule to suit your local conditions. By keeping detailed planting and harvesting records over time, you can gradually develop a fine-tuned schedule that suits your farm location and mix of greens. Your state's Cooperative Extension vegetable specialist may also have a sowing schedule that can be adapted.

Dutch growers space lettuces at 15 to 20 plants per square yard. The growing period determines the actual density. A 12" x 12" spacing equates to approximately 16 plants per square yard, while a 10" x 10" spacing is 20 plants per square yard. The tighter spacing is used in the slower growing period of winter. Equal spacing between plants and between rows produces more uniform heads.

For production on a small scale, a lawn roller, with studs attached at appropriate spacing, will both mark the correct spacing and punch transplant holes in the soil. It will save a lot of time, both at planting and at cultivation time. In general, lettuce plants held as seedlings for more than three to four weeks take a relatively long time to mature and don't produce good heads. Seedlings that might appear too young to transplant often grow very quickly and make beautiful heads. Considerable field experience suggests it is wise to transplant a small, healthy, hearty plant.

Deeper, less-frequent waterings are preferred to lighter, more-frequent irrigation. Spot watering may be necessary in places (such as the edges of beds) that dry out quickly. The crop should not suffer for water, but it is also important to keep the soil surface from being wet for long periods of time. This reduces disease pressure.

Growing and Transplanting Lettuce Seedlings

If you want to try growing lettuce transplants at home, you'll need a cold frame, greenhouse, or a hotbed. In a cold frame, you'll need about 11 weeks to grow your seedlings so that they're ready to plant in the garden. If you use a hotbed or a greenhouse, your seedlings may mature as quickly as in a month. Plant the seeds about 5 inches apart and as they grow you can thin them to about 2 ½ apart so that they have plenty of room to grow. Providing ample space for your seedlings also helps avoid problems with fungal diseases. When you're ready to transplant your lettuce, you should get it used to outside conditions. About a week before you're ready to plant, set your plants outside a few hours, increasing the time to a full day after about 3 days. Your lettuce should be ready to plant after a week of exposure to outside temperatures. Wet down your garden beds several hours or a day before you transplant.

Harvesting

Harvesting your lettuce at the correct time is important for maximising its flavor and texture. If you let your lettuce crop sit in the garden too long, it will increase in bitterness and the leaves will become tougher. Lettuce is generally ready to harvest about 80 days after planting with seeds and about 60 days after using seedlings. If you grow romaine or butter leaf lettuce, you should first remove the outer leaves, then dig up the entire plant and cut it off at the base. This way, it's possible to have a second crop. With leaf lettuce, simply remove the outer leaves and let the plant continue to grow.

Small-scale production usually entails laborious hand-cutting of greens using a sharp knife. In recent years, however, some small greens harvesters have come onto the market. Unfortunately, small-scale salad harvesting equipment is still a rather new phenomenon—especially in this country —and very little information has been published to date.

Post-harvest Handling

While specialty lettuces command higher prices than ordinary iceberg, they also demand specialty handling. They should be immersed in near-freezing water almost immediately after harvest. Rapid removal of field heat is essential to maintaining product quality, and is a key element of professional handling. Depending on how the greens are cut, it may be helpful to pack any free leaves or smaller plants loosely into a mesh sack to keep them "corralled" during the hydro-cooling process. The University of Wisconsin has developed an excellent fact sheet on the use of mesh produce bags for this purpose.

Another critical (and demanding) component of handling is careful and thorough washing, followed by equally careful and thorough sorting. Greens are almost always eaten raw, and many consumers do not wish to go to the trouble of re-washing or picking through the evening's salad. Most professional lettuce growers will say that few things drive customers away faster than dirty, gritty, or poor-quality greens. Once the greens are well washed, gently drained dry, and ruthlessly graded, they should be packed into their final retail container (one that allows a bit of breathing), and returned immediately to a near-freezing high-humidity (95%) environment. They should be kept under those conditions until delivery to the retailer or the final consumer.

Commonly, greens are harvested from individual beds and are mixed after harvest. One mesclun grower advised extra care be taken at this stage since the tender leaves are very susceptible to bruising during the mixing process. He only mixes the greens as he bags them, after the washing and drying process. After cutting, the salad greens need to be pre-washed and dried for best post-harvest quality. A restaurant- or industrial-scale salad spinner is an efficient machine for washing and drying the greens.

LETTUCE PESTS AND DISEASES

Lettuce is affected by a number of pests. Be especially on the lookout for cabbage worms, loopers, aphids, and armyworms. Lettuce that is harvested in the spring has fewer problems with diseases. However, fall and summer crops may attract a number of pests as pest larvae have had a chance to develop during the warmer weather.

As far as diseases, fungal diseases are sometimes a risk for your lettuce crop. Look for problems with damping off, sclerotinia, and mildews. To avoid these problems, use an organic fungicide, and avoid over watering. A good way to deal with over watering is to use a drip irrigation system or a soaker hose.

Bolting is a common problem with lettuce that causes flowers and seeds to grow and the lettuce to taste slightly bitter. It is caused by hot and sunny conditions. It affects many plants in the cabbage family as well. You can also use companion planting for your lettuce to reduce problems with bolting. Sweet corn makes a good companion plant for lettuce as it will shade the plants during the hottest months of summer. Read more about bolting at this website from Iowa State University.

Pests

Luckily, lettuce is attacked by relatively few insect pests, primarily aphids and thrips. Both are fairly well controlled with an insecticidal soap, such as M-Pede™ brand (also sold to gardeners as the Safer™ brand). If the water is hard, the spray solution should be mixed with softened or distilled water because the calcium in hard water sequesters the fatty acids and greatly reduces the potency of the spray.

Young plants are sometimes subject to pressure from flea beetles. Arugula seems especially attractive to this pest. The most effective non-chemical control is a light row cover such as Agribon+™ AG-12 insect barrier, which can exclude most pests while providing a more sheltered microclimate. Cutworms can also be a problem. If present in the soil, the best non-chemical approach may be to hunt them down (dig in the soil near affected plants) and crush them. Cutworms curl up when dug out of the soil Insect and disease pest problems in greenhouse production are similar to those of field production, yet unique because of the modified enclosed environment inside a greenhouse. Fortunately, great strides have been made in recent years in developing effective greenhouse IPM programs. European and Canadian researchers have developed the use of many control tactics for greenhouse pests.

Integrated pest management (IPM) programs monitor pest levels with such tools as pheromone traps, yellow sticky cards, and on-site examination of plants with a hand lens. Buildup of pest populations can be detected and treated before they reach injurious levels. Cultural methods may be altered to make the environment less favorable for particular pests. Physical barriers such as insect screens may be used to prevent some types of pests from entering a greenhouse. It is helpful to remove infested material from the greenhouse to physically reduce pest populations. Screening air intakes to the greenhouse is also effective—especially as a means to control thrips. Several natural predators can be obtained and released to reduce aphid and thrips numbers to acceptable levels.

Common Insects that Affect Lettuce

The following is a list of common insects that attack lettuce and how to deal with them. Please refer to this excellent website from Arizona University for more information and some great photos of common pests that affect lettuce crops.

Aphids

Aphids are tiny insects that will cause wilting on the plant. You can look for a sticky fluid on the leaves commonly called "honeydew." The insects vary in color from yellowish to light green. Aphids also carry diseases that may affect your lettuce, so definitely deal with them as soon as you see evidence of their presence. Try blasting your plants with a high pressure hose, or use an organic insecticide. Also, keep your garden weed free, as aphids can make their homes in weeds as well.

Cutworms and Armyworms including Amanthes c-nigrum, and Spodoptera praefica

These insects will eat the stems and leaves of lettuce. Armyworms vary in color from brown to green and have black spots. Cutworms are a little lighter in color. Try picking them off by hand or use an organic pesticide.

Loopers

Loopers that attack lettuce include the Alfalfa Looper, Autographa californica and the Cabbage Looper, Trichoplusia ni. These are dark green worms that are quite thin. Look for dark stripes. They are called "loopers" as they crawl by making a loop, much like an inch worm. Organic pesticides with Rotenone are excellent organic pesticides that will work on your lettuce to kill loopers.

Tarnished Plant Bug (Lygus lineolaris)

This is one of the most common garden pests in the U.S. This pest eats the head of your lettuce. They are oval shaped and have black, yellow, and brown markings.

Western Spotted Cucumber Beetle (Diabrotica undecimpunctata)

These pests are more a problem for seedlings than for adult lettuce plants. They are yellow and have black spots. They usually get no larger than 8 millimeters long. They are also hosts to a number of diseases including bacterial wilt and cucumber mosaic virus. Lettuce is especially at risk when you grow cucumbers in the same vegetable plot. Read more about this pest at the National Sustainable Agriculture website.

Diseases

While few diseases other than damping-off affect *young* lettuces and greens, older plants face more challenges. The limited number of fungicides and other disease control agents available to organic growers makes disease prevention crucial in organic production. The cool moist conditions typical of fall and spring production may encourage diseases in the field. *Sclerotinia* drop is caused by fungi that attack the lower leaves in contact with the soil, producing a cottony growth. Eventually the entire plant may "collapse," with the leaves dropping down on one another in succession. Bottom rot, caused by *Rhizoctonia* species, is another potentially serious soil-borne fungal disease of cool-season lettuce. It occurs on nearly mature plants, first appearing as slightly-sunken rustcolored lesions (plus or minus some amber ooze) of varying size. Affected plants will have a very short shelf life if harvested, and may decompose into a slimy black mass.

New biofungicides have been developed that can assist in disease management. An example is SoilGard™, a product of Thermo Trilogy, Inc. It contains the naturally-occurring fungus *Gliocadium virens*, which can be used to control disease-causing *Fusarium*, *Pythium*, *Rhizoctonia*, and *Sclerotinia* fungi.

Foliar sprays of compost- and plant-extracts (e.g., horsetail, stinging nettle, valerian), are being tested for their use as fungal-disease suppressants. To test the disease-suppressive potential of compost watery extracts, researchers in Germany combined one part well-rotted manure-based compost to six parts water and let it sit for one week. It was then filtered and used as a spray. Applied every five to ten days, the extract prevented powdery mildew, downy mildew, late blight, botrytis blight, and anthracnose on tomatoes, beans, potatoes, grapes, and sugar beets. It was also effective in preventing damping-off diseases.

A disease (actually, "physiological disorder") of particular concern on lettuce is tipburn. Tipburn occurs when a sudden change in weather causes more rapid transpiration than water uptake. Symptoms include browning and rotting of the edges of internal leaves, which may not be visible from the outside of the head. This disorder is related to calcium deficiency and is aggravated by high soil fertility and high temperatures. On greenhouse lettuce it is particularly a problem during sunny early-spring days; in the field, it can be a problem if irrigation is mismanaged during hot weather. Assurance of adequate calcium and avoidance of excess nitrogen and potassium will work to minimise problems. Minimal venting (thus decreased transpiration) during the critical period is also important for greenhouse-grown lettuces. Shading may also be used to reduce transpiration, as may watering by mist or spray as long as it doesn't compromise disease control. Acid soils and high soil salt content increase risk.

Common Diseases of Lettuce

Here's a list of common diseases and other problems that can affect lettuce and how to deal with them. You can read more about lettuce diseases at this University of Arizona website.

Anthracnose (Microdochium panattonianum)

This fungal disease can survive in garden debris of infected plants or in the soil. Therefore, it's best to destroy infected plants as soon as possible as the disease can be present even if the plant is dead. Don't compost infected plants!

The disease can remain present in the soil for several years. Extremely moist conditions will attract the fungus. Keep foliage dry when possible. There are lettuce varieties that are resistant to the disease, so ask at your local nursery for disease resistant seeds if you live in an area of the country with a lot of rainfall.

Bolting

Bolting is a common problem with lettuce that causes flowers and seeds to grow and the lettuce to taste slightly bitter. Bolting is caused by hot and sunny conditions. It affects many plants in the cabbage family as well. You can use companion planting for your lettuce to reduce problems with bolting. Sweet corn makes a good companion plant for lettuce as it will shade the plants during the hottest months of summer. Read more about bolting at this website from Iowa State University.

Bottom rot (Rhizoctonia solani)

This is a fungus that attacks mature lettuce in poorly-draining soil. Look for areas of necrosis on both the stems and on lower leaves. In other words, keep an eye out for brown, dying spots. The lettuce will also become wilted and start to decay. Add compost to your soil before you plant lettuce to ensure proper drainage. Try using a soaker hose or drip irrigation to avoid over watering.

Damping off (Pythium ultimum and Rhizoctonia solani)

Damping off is another soil-based fungus. This disease is mostly a concern for your young lettuce plants. As the fungus lives in the soil, the disease can start to take effect when your seedlings are still in the ground. Look for water-filled spots and plant wilting as the lettuce starts to mature. Damping off occurs most frequently when conditions are wet and damp. Make sure you have well draining soil, and try using drip irrigation or a soaker hose to effectively irrigate your plants.

Downy mildew (Bremia lactucae)

This disease gets its name from the downy patches that form on the underside of the leaf as the disease matures. Another telltale sign is yellowing on the upper leaf surface. The disease is fungal, and when you see the downy deposits, it will soon spread by spores.

Once infected, the lettuce will eventually turn brown and die. Keep your garden weed free as this disease can live on weeds around and in your garden, even through winter. Moist conditions exacerbate the problem, so try using drip irrigation or a soaker hose to keep the foliage dry. Look for disease resistant varieties of lettuce to further combat this disease. Try out a organic fungicide to treat problems with downy mildew as well.

Mosaic virus

Aphids can play a role in transmitting this virus. Do your best to keep aphids off your lettuce. When young plants are affected by the mosaic virus, the leaves will start to curl and their will be a light brown discoloration along parts of the leaf. Keep your garden weed free to reduce habitat for aphids, and control aphids as soon as they appear. Remove plants that show signs of this disease immediately and destroy them. Try out a natural Neem Seed Oil pesticide to get rid of aphids.

Varnish spot (Pseudomonas cichorii)

Varnish Spot is a bacterial disease found in the soil. This is a pretty uncommon disease for most home gardeners, so you're unlikely to have problems with it. Look for dark brown spots on the leaves. It usually attacks on the inner leaves of lettuce, making it hard to spot. Frequent crop rotation can help avoid this problem.

Watery soft rot (Sclerotinia sclerotiorum)

This disease is present in the soil and can survive the winter on garden debris from your lettuce crop. All the more reason to do a thorough cleaning of your garden once you've harvested your crop! This disease affects young plants especially. Look for rotting, water-filled tissue on your lettuce plants. This disease is mostly present in cooler, wet weather.

Weeds

Many greens crops are poor competitors with weeds. Additionally, weeds growing within rows can inadvertantly be harvested along with the crop and contaminate the product. As a result, considerable effort may be expended in cultivation and handweeding to assure a clean field. Crop rotations that include careful selection of cover crops can go a long way in reducing weeds in vegetable crops. In some instances, the residues of cover crops are left in place and the crop is seeded or transplanted into it. The residue mulch conserves moisture, reduces erosion, and may suppress weeds through smothering and through allelopathy. The sensitivity of small-seeded crops like lettuce, however, may require extra pains in making such systems work well.

REFERENCES

Molly Damon, Nancy Z. Zhang, David B. Haytowitz, Sarah L. Booth (2005). "Phylloquinone (vitamin K1) content of vegetables". *Journal of Food Composition and Analysis* (Elsevier) **18**: 751–758.

Simpson, D.P. (1979). *Cassell's Latin Dictionary* (5 ed.). London: Cassell Ltd.. pp. 883.

Ashworth, Suzanne (2002). *Seed to Seed*. Seed Savers Exchange.

Wim J. M. Koopman, Eli Guetta, Clemens C. M. van de Wiel, Ben Vosman and Ronald G. van den Berg (1998). "Phylogenetic relationships among Lactuca (Asteraceae) species and related genera based on ITS-1 DNA sequences". 1517–1530.

Grigson, Jane (1978). *The Vegetable Book*. London: Penguin.

14

Potato

Potato is the world's fourth important food crop after wheat, rice and maize because of its great yield potential and high nutritive value. It constitutes nearly half of the worlds annual output of all root and tuber crops. With an annual global production of about 300 million tonnes, potato is an economically important staple crop in both developed and developing countries.

In India, more than 80% of the potato crop is raised in the winter season (Rabi) under assured irrigation during short winter days from October to March. About 8% area lies in the hills during long summer days from April to October. Rainy season (Kharif) potato production is taken in Karnataka, Maharashtra, HP, J&K and Uttranchal. In major parts of China (which is one of the leading producers) potato is harvested in the month of October to December and in few regions it is harvested in month of May -July.

PRODUCTION TRENDS

Potatoes are grown in about 150 countries throughout the world and more than a billion people worldwide eat potato. About 328.87 million tonnes of potato are produced in the world over an area of about 19.13 million hectare.Area under potato cultivation is almost constant and same is the case with production. However, there is a slight decline in the yield. The top ten producers in the world are China, Russia, USA, Ukraine, Poland, Germany, Belarus, Netherlands and France. These together contribute about 70% of the total production. India ranks third contributing around 7.5% to the world's production. Production wise India has always remained in the top ten since last twenty years. Europe is the largest per capita consumer, followed by North America and Latin America. However, there is a slight decrease in area under cultivation in China but there is increment in production. Same is the case with other major countries.

India ranks 4th in area and it is the 3rd largest country in world in production of potato after China and Russian Federation. Potato is produced in an area of 14.00 lakh ha with a production of 250 lakh tonnes and productivity of 17.86 ton per ha.

Production of potato has remained almost stagnant around 25 million tonnes since 1999 - 2000. Production declined from 25 million tonnes in 1999-2000 to 24.44 million tonnes in 2000-2001 and 2001- 2002 and increased to 26.5 million tonnes in 2002-2003. Potato production declined again to 24.5 million tonnes in 2003-04. In 2004-05 there is an increase in the production to about 25 million tonnes.

Potato is grown almo st in all states of India. Major potato growing states are Himachal Pradesh, Punjab, Uttar Pradesh, Madhya Pradesh, Gujarat, Maharashtra, Karnataka, West Bengal, Bihar and Assam. UP, West Bengal, Bihar and Punjab together account for about 86% of India's production. However, potato consumption per capita in India (14.8 kg/head/year) is one of the lowest in the world and hardly 1% of the potato is processed. 90% of the potato crop in India is cultivated on Indo-Gangetic plain from October until February-March.

The major countries to which India exports are Nepal followed by Srilanka, Mauritious, Maldives and USA. Nepal shares almost 82%of total exported quantity. Liberalisation was having a great impact on the trade of potatoes as free trade was allowed after that. The pre liberalisation and post liberalisation trends are quite clear. After liberalisation there was almost a four-fold increase in the export. Not only there was an increase in quantity but there was also an increase in terms of value.

India basically imports starch from other countries. Starch is used in noodles and other value added food items. Since 2004-05 the starch import has come down because some Indian industries have also started manufacturing potato starch. The government is also planning to set up a potato starch unit in Faizabad (UP).

Over the years price has remain at the top during October -November. This is because of lean season with limited availability of supplies from cold storages during these months. The only source is produce from Karnataka which is not able to meet the high demand. Most of the harvest starts November onwards. As more and more produce start arriving in the market prices start coming down. This is evident from December to April when price has remained lower. By April harvesting stops and Potato from cold storage is utilised and by October cold storage is almost empty. This trend is almost similar in all the major potato markets in India.

Generally production of potatoes is about 25 million tones which is the normal requirement also for the consumption including requirement for seed bulbs, processing industries, export, waste and storage losses etc. Present estimated production is about 24.51 million tones which is little less to meet the demand, which could be compensated by the restrictions on export and also by increasing area under kharif potato production. Presently, In Agra region almost 45-50% of the stocks are already used and only 30% is there for seed purpose. Altogether, taking into account Agra and near by area, stocks remained for coming months for consumption purpose is only 30-35%.

Prices in the month of Aug-Sep will be governed by arrivals from Hasan market (District in Karnataka) which will start from 15th Aug. Rain in Hassan which is one of the major producing centers will also have impact on the potatoes prices. For this year crop is expected to be in good condition and there will be increase in supply in southern market which will pressurise the prices. The produce from Hasan will cater the market till the end of September after which in the month of Oct – Nov there will be again dearth of supply and prices will go up till November. In Northern region, harvesting will start in the month of Nov and fresh Rabi produce will come in the market which will drive the market down the months of December and January.

CULTIVATION

Climatic Requirements

Potato basically requires relatively mild temperature during early growth and cool weather during tuber development. It does well under well-distributed rains or moist weather situations to high temperature, humidity rains are not conducive to potatoes as these lead to insect-pest, disease, viruses epidemics. Impeded drainage or lack of aeration also is considered harmful as it restricts the tuber development. Therefore, climatic conditions are very crucial for determining the potato yield. Its general and specific climatic requirements are delineated as under.

Potato is a temperate or cool season crop which needs a low temperature, lower humidity, less windy and bright sunny days. It thrives best in cool regions where there is sufficient moisture and fertile soil. Potato needs about 25°C at the time of germination, about 20°C for vegetative growth but between 17-20°C for tuberisation and tuber development. It is noticed that higher temperature has an adverse effect on the tuber growth, where as temperature above 30°C stops tuber formation completely. It is probably because at higher temperature the rate of respiration increases and the carbohydrate formed by the process of photosynthesis is consumed rather than stored in the tubers. Under higher temperature the plants are elongated leaves become wrinkled and plants have silky appearance. Tubers do not develop under such situations. Cloud days, rains and high humidity are very congenial for spread of fungal, viral and bacterial diseases, and such conditions are not good for the crop.

Potato is a crop of temperate climates and its importance is great both for human consumption and for industrial uses. There are different varieties of potatoes which are classified on the basis of the length of the vegetative period. There are early varieties with 90 days vegetative period; others are medium verities with 90-120 days of vegetative period and still others are late varieties with over 120 days of vegetative periods.

Figure 1. Potato plant

Temperature

Potatoes are planted when the soil temperature at a depth of 4 inches is 8 ° to 10 °C. When they are planted at a higher temperature, the tubers are damaged, but when they are planted at a lower temperature, the sprouting is retarded. It has been estimated that a mean daily air temperature of 10 ° C, the length of the period from sowing to sprouting is 27 days. But when the mean daily air temperature is 22 ° C, sprout appears after 12 days. Experiments have shown that a most favourable temperature for the potato tubes is 18 ° to 20 ° C. At 22 ° to 25 ° C the formation of the tuber is retarded and beyond 29 ° C it completely stops. Mild frost to the extent of -1° C partially damages the potato leaves, but when the temperature falls below -2° C, the exposed parts of the most of the varieties are destroyed.

Moisture

The water requirements of the potato crop during its vegetative period are not very high. It has been estimated that rainfall of 15 to 20 cms or an equivalent amount of irrigation, is sufficient for the normal crop. At the time of germination, the crop requires little external moisture, because the potato tubers themselves contain adequate amount of water. However, as the leaves grow after sprouting, the water requirements of the crop increase. The optimum soil moisture for the potato crop is 65 to 80 percent at the field capacity. Just after sowing and before sprouting, heavy rains adversely affect the crop. Alternate dry and rainy weather leads to an excessive growth of potato plant and encourages the formation of the off-shoots in the young tubers. Potato requires a high light intensity if the crop remains in shade or if it is too densely planted, the yield is adversely affected.

Land Preparation

Potato being a tuber crop, requires a soft, friable and deep seed bed for the development of uniform large smooth tubers. To provide this, potato often is planted on raised seedbeds / ridges, prepared with the help of potato ridger after a thorough and deep cultivation of soil, liberally fertilised with F.Y.M., compost or other organic manures including the green manures.

All the stubbles and non-decomposed organic matter is removed if it is feared that it may not be decompose quickly. Every care is taken is to eradicate the weeds. Potato in Punjab usually follows a Kharif fodder or grain crop of maize or rice which are not only well fertilised but also leave sizeable quantities of root-mass after their harvest thus enriching the soil.

There is little time left for thorough incorporation of organic manures, its decomposition and land preparation. In true sense, speedy land preparation to ensure timely planting of this short duration crop holds the key to profitability margins realised.

If the crop is sown early, it is likely to attain marketable maturity early in the season when the prices are comparatively high and availability of fresh and quality tubers is short. Potato harvested at this time of the year thus sells at 4 to 5 times the peak production period rates and even if the yields are some what lesser than the optimum yet the farmers end-up realising 2-3 time more income than the returns realised from peak-season harvests.

Moreover an early harvest also means an early vacation of the field which in turn means an early sowing of following wheat (the most preferred sequence crop to potato). Or early sowing of the next potato or vegetable crop in early December instead of delayed sowing in end December / first fortnight of January.

Wheat is highly sensitive to delayed planting and delayed sowings. Its grain yield tends to decline @ 30-400 kg/ha/week and therefore 2-3 weeks gain in planting time of succeeding wheat transforms to 7-10 quintal gain in the yield of wheat (Rs 4000-6000 more returns/ha).

Early planting of potato in the season thus proves doubly gainful and hence is to be practiced atleast over 25-30 percent of the area, if not more, to take advantage early high market prices.

Thus available time becomes a limiting factor and its efficient management deserves special attention in the management of potato cultivation, especially the land preparation tillage and planting of the crop.

Land preparation tillage in Punjab is largely accomplished with the help of tractor operated implements, mostly through repeated use of off-set disc harrow in the initial stages followed by 2-3 cultivator operations and planking given cross-wise to break the bigger clods and pulverise them into small size aggregates.

Usually, the first one half dose of N and full dose P and K fertilisers is broad cast at the time of last cultivator operation and mixed thoroughly in top 2-3 inches of soil by giving a light harrowing.

This is followed by planting operation wherein potato tubers are manually placed in the rows, 20 cm apart, with a row to row spacing maintained at 60 cm. Soon after, a ridger is run to cover the potato tubers by throwing the soil from both the sides and ridges pressed and the first light irrigation applied (by allowing the filling to about ½-2/3rd of the ridge height).

Use of Freshly Produced Tubers for Seed

- The potato tuber raised from autumn crop can be utilised can be utilised for spring crop sowing after breaking the dormancy of produce.
- While breaking dormancy, the following points must be taken into account.
- Select the produce for seed purpose by separating the tubers to be used for tubers.
- Keep the produce free from soil adhering on its surface by arranging water washing.
- On account of cold weather during spring sowing, the seed tubers may be cut in to small pieces.
- While giving cuts to tubers, eyes should not be damaged.
- Each tuber can be divided in to 2 - 4 pieces, and take care to cut the longitudinally as there are more eyes on the crown end than at the bottom.
- Each cut piece should have the weight of 30 g.
- Each cut piece should have at least two eyes on its surface.
- Do not sow the cut pieces without breaking dormancy. The autumn crop tubers have about 2 - 3 months dormancy period

- Use Ethylene Chlorohydrin (1 liter for 2 tons of seed tubers) followed by the thio-urea solution (1% or 1 kg of chemical for 1 ton of seed tubers) followed by final treatment with GA (Gibberellic acid 1 ppm for 10 minutes).
- Treat tubers with Ethylene Chlorohydrin (without making into pieces) for 24 hours, @ 1 liter for 2.0 tons of seed under cover or in an enclosed space.
- For effective use of Ethylene Chlorohydrin, dig-out trenches and put the seed in it and treat with this chemical taking care that these trenches when closed are air proof.

Follow the above treated material with thio-urea treatment.

- Dissolve 1 kg thio-urea in 100 liter of water and dip the seed in lots for one hour.
- Solution made once can be utilised for the other lots and seed tubers should be free from soil.
- The same solution can be utilised for 5 - 6 lots without affecting the efficiency of dormancy breaker.
- Prepare separately, 1ppm solution of Gibberellic acid (1 mg in 1000 liter water) and keep the tubers in this solution for 10 minutes
- Cut the seed into small pieces for sowing spring crop of potato.

Germination Capacity

- Maximum germination power will be in 12 months after harvest and then decreased.
- When the tuber is planted before and after the proper age, °eneration is proportionate to the time interval.
- Seed tubers of the longer storage grow more stem than those from short storage, number of stems corresponding to age of the tubers
- Germinating potato tubers at 20 °C resulted in a morbid yield increase over those sprouted at 10 ° C before planting.
- Germinating at 20 ° C in the dark also increased branching of sprouts and no. of stems. Under these conditions gibberellin activity in leaves decreased and that in sprouts increased.
- The low temperature germination increased the gibberellin and auxin activity in leaves.

Seed Sprouting

- Weight of sprouts increases with age of tubers as a result of accelerated sprout growth and an increase in the number of sprouts per tuber

– Age of tuber may be detected by changes in type of sprout growth

1. The one sprout stage
2. The multiple sprout stage
3. The branching stage
4. The small tuber formation stage.

– As number of stem increases, the number of tubers formed also increases and average size of tuber is likely to decrease. Sprouted seed with only 1 or 2 sprout per piece result in a crop with a higher proportion of large potatoes under these situation.

– When all but one sprout were removed before planting, an average of 20.5 tuber per plant were produced, many of saleable size

– When seed pieces were restricted to one stem the no. of tubers was reduced and high proportion of large tubers were formed.

– Removal of the first sprout results in a greater number of sprouts to develop.

– The no. of sprouts increases in proportion to the weight of the seed piece.

– Seed stored under warm conditions soon after harvest produce few sprouts

– Pre germination of tubers results in a greater development of the root system and a higher proportion of root to tops. Early varieties react more in this manner than do late ones.

Method of Planting

The main factors considered for method of planting are

1. Purpose for which Potatoes are grown
2. Area to be planted and size of field.
3. Availability of tractor and Pota · planter.
4. Availability of labor.
5. Availability of seed size plantir g

The plant spacing depends upon the purpose for which the crop is sown i.e. whether for ware / commercial purpose or seed purpose.

1. If the crop is sown for seed purpose and large sized seeds are sown at narrow spacing, it will produce large number uniform sized tubers.
2. If the large sized potatoes are sown at wider spacing, it will produce large sized and

fewer tubers generally, medium sized tubers (3.75 cm diameters) are considered suitable for Potato sowing.

— If large area to be planted, potato planter not only save labor but also ensure timely and accurate sowing at desired depth and spacing.

— In case of small area, sowing with spades prove equally beneficial.

— Moreover, in small fields where heavy machinery does not hold good, manually sowing is preferred.

— On account of the use of potato planter seasonally, small and marginal farmers do not afford to purchase it. However, large farmer do prefer the sowing with planter.

— Availability of labor and as and when required also decide the sowing.

— No labor scarcity at cheaper rate, sowing manually proves equally efficient.

— The spacing between the intra-seed placements depends upon the seed size.

— Large and medium sized Potato for attaining high tuber yield should be placed at a distance of 15-20 cm.

— The distance between row to row is recommended for high tuber yield realization at 60 cm apart.

For manually planting of tubers, following points need attention

1. Apply all fertilizer at the prepared field by broadcasting which will be mixed with soil at ridge formation with spades / manual bund maker (Jandra).
2. Place the Potato tubers at the marker sign at a fixed distance.
3. The soil is placed on these tubers up to 7.5 cm height with manual bend maker (Jandra) or with spade.
4. Too high ridges formation delays the germination of tubers.
5. Less than 6 cm height of ridge cause sun-burn to the tubers.
6. All the Potato tubers should be covered thoroughly.
7. Care is taken that seed does not come in direct contact with fertiliser otherwise the seed will not germinate.
8. Apply irrigation after the sowing of Potato tubers, to provide sufficient moisture required for the germination. Of the tubers and repeat irrigation after 8 - 10 days interval.
9. Do not allow the level of water to rise more than half otherwise the Potato tubers will rot, due to crustation of top soil.

10. Flat planting is common in areas having light sandy soils.
11. Do follow 2 earthing; first 25-30 days after sowing and second 40-50 days after sowing in flat planting it. This method is not considered a scientific one.
12. Planting on ridges is done after preparation of field; the ridges of 30 cm height and 60 cm width are made with the help of spade (Jandra). Planting is done either at the top or on the sides of ridges.
13. Scientifically, Planting with Potato planter has been found very effective with respect to the germination and for maintaining the proper pant population stands.

Crop Rotations And Intercropping

Potato is grown in rotation with other crops, usually to maintain:

- Desirable soils texture and state of fertility
- To build up organic matter
- To reduce crop loss from insect and plant disease
- To increase per unit productivity per unit area and time
- To improve the quality of produce

Potato, being a fast growing crop, fits well in different rotations and inter cropping systems. It can very successfully be grown in the following sequences.

1. Maize - Potato - Wheat
2. Paddy - Potato - Wheat
3. GM - Potato - Wheat

For successful cultivation of these rotations, the following time schedule needs to be considered.

- Maize and paddy sowing should be done in early June
- Grow short duration early maturing varieties
- Grow potato in October
- Prefer late sown varieties of wheat viz. PBW373, PBW138, Raj 3765 in the end of December and early January
- Green manuring is done during rainy season
- Green manure should be ploughed in the field at least 15 days before potato planting
- Green manure improves the soil fertility

The potato crop is also grown in different inter-croppings such as

1. Sugarcane + Potato
2. Maize + Potato

Seed and Seed Treatment

- Potato is propagated vegetatively.
- Hence the disease pathogens are carried from mother plants and the crop gets spoiled.
- Therefore the seed must be produced by following seed production techniques.
- Seed should be pure, healthy and of uniform size for getting high yield.
- Following points need to be considered while making selection of potato for seed purpose.
- Seed tuber should be uniform in shape, desired size and colour
- The ideal tuber size for planting should be 3.5 - 5.0 cm in diameter and should be graded
- Seed tubers should be purchased from the trusted agency like National Seed Corporation or State Seed Corporation or State Department of Horticulture or CPRI
- Seed tuber should not be shriveled
- Seed tuber showing, black scruff, potato scab, brown rot and nematode infection must be separated before planting.
- Keep the seed in diffused sunlight under shade and do not sow the seed directly from the cold storage.
- Allow the seed to acclimatise with the normal conditions and to develop the sprouts in shade.
- The weight of each tubers may vary from 30 - 50 gm.
- Too small seed (chats) must be rejected ruthlessly as they often are virus infected.
- The seed is considered suitable for sowing when the sprout becomes 6-8 mm long.
- While collecting seed for sowing purpose, handle the sprouted seed with care and gently so that sprouts do not break / detach.
- The weight of each tuber may vary from 30 - 50 gm.
- Too small seed (chats) must be rejected ruthlessly as they often are virus infected.

Following points are to be considered while giving seed treatment to tubers

- The tubers should be free from soil because soil adhering to potato reduces the efficiency of fungicide
- Treat the tubers even when taken from the cold storage or from any other source
- Treatment should be given before the sprouting initiates
- Select the fungicide and adjust its dose accordingly
- Common approach in Punjab is to treat seed with Agallol 0.5% or Aretan/Tafasan 0.25% solution prepared by dissolving 500 g of Agallol or 250g of Tafasan/Aretan in 100 liters water

USES OF POTATO

In India, processed potatoes currently constitute less than 0.5% of annual production. While the consumption of processed potato products is anticipated to increase, at present, the processing sector is largely comprised of various kinds of dehydrated potato products, starch etc.

Netherlands (ninth in production and the highest exporter and importer) exports around 22-28% of its production. Whereas, India being the third largest producer exports hardly 0.45%. This is because Netherlands has the comparative advantage of weather that allows it to grow varieties that are suitable for further value addition like chips and French fries. The same is the case with Germany and Belgium. But in India the climate required to grow varieties for French fries and chips is not so condusive. But now some varieties like ChipsonaI and ChipsonaII are being developed by ICAR especially. Once these varieties are fully developed they will give a good competition to the European value added potato varieties.

The world's import and export price have gone hand in hand. After 1995 both the import and export prices of world had started falling. But India's export and import prices have never been close to each other. There is a significant gap between India and world's price. But both the price has a direct correlation. With rise in world's export price India's export price has also increased. After liberalisation India's import price has also moved closer to world import price.

REFERENCES

Larry Zuckerman (1999). *Potato, The: How the Humble Spud Rescued the Western World*. Douglas & McIntyre.

Lang, James (2001). *Notes of a Potato Watcher*, Texas A&M University, College Station, TX.

Salaman, Redcliffe N. (1989). *The History and Social Influence of the Potato*, Cambridge University Press (originally published in 1949; reprinted 1985 with new introduction and corrections by J.G. Hawkes).

Hawkes, J.G. (1990). *The Potato: Evolution, Biodiversity & Genetic Resources*, Smithsonian Institution Press, Washington, D.C.

15

Tomato

The tomato is grown worldwide for its edible fruits, with thousands of cultivars having been selected with varying fruit types, and for optimum growth in differing growing conditions. Cultivated tomatoes vary in size from tomberries, about 5mm in diameter, through cherry tomatoes, about the same 1–2 centimetres (0.4–0.8 in) size as the wild tomato, up to "beefsteak" tomatoes 10 centimetres (4 in) or more in diameter. The most widely grown commercial tomatoes tend to be in the 5–6 centimetres (2.0–2.4 in) diameter range. Most cultivars produce red fruit; but a number of cultivars with yellow, orange, pink, purple, green, black, or white fruit are also available. Multicolored and striped fruit can also be quite striking. Tomatoes grown for canning and sauces are often elongated, 7–9 centimetres (3–4 in) long and 4–5 centimetres (1.6–2.0 in) diameter; they are known as plum tomatoes, and have a lower water content. Roma-type tomatoes are important cultivars in the Sacramento Valley where a 120-acre Morning Star cannery handles 1.2 million pounds of tomatoes an hour during the harvest season where the fields yield about 40 tons to the acre.

Tomatoes are one of the most common garden fruits in the United States and, along with zucchini, have a reputation for outproducing the needs of the grower. As in most sectors of agriculture, there is increasing demand in developed countries for organic tomatoes, as well as heirloom tomatoes, to make up for flavor and texture faults in commercial tomatoes. Quite a few seed merc' 's and banks provide a large selection of heirloom seeds. Tomato seeds are occasionally organically produced as well, but only a small percentage of organic crop area is grown with organic seed. The definition of an heirloom tomato is vague, but unlike commercial hybrids, all are self-pollinators who have bred true for 40 years or more.

VARIETIES

There are a great many (around 7500) tomato varieties grown for various purposes. Heirloom tomatoes are becoming increasingly popular, particularly among home gardeners and organic producers, since they tend to produce more interesting and flavorful crops at the cost of disease resistance, and productivity. Hybrid plants remain common, since they tend to be heavier producers and sometimes combine unusual characteristics of heirloom tomatoes with the ruggedness of conventional commercial tomatoes.

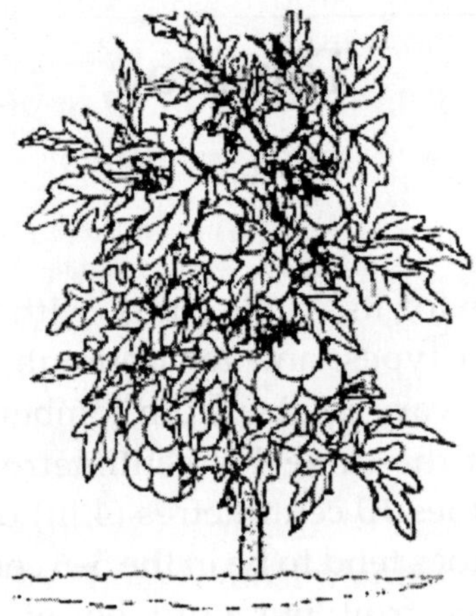

Fiure 1. Tomato plant

Tomato varieties are roughly divided into several categories, based mostly on shape and size. "Slicing" or "globe" tomatoes are the usual tomatoes of commerce; beefsteak are large tomatoes often used for sandwiches and similar applications - their kidney-bean shape makes commercial use impractical along with a thinner skin and being not bred for a long shelf life; globe tomatoes are of the category of canners used for a wide variety of processing and fresh eating; oxheart tomatoes can range in size up to beefsteaks, and are shaped like large strawberries; plum tomatoes, or paste tomatoes (including pear tomatoes), are bred with a higher solid content for use in tomato sauce and paste and are usually oblong; pear tomatoes are obviously pear shaped and based upon the San Marzano types for a richer gourmet paste; cherry tomatoes are small and round, often sweet tomatoes generally eaten whole in salads; and grape tomatoes which are a more recent introduction are smaller and oblong used in salads. Early tomatoes and cool-summer tomatoes bear fruit even where nights are cool, which usually discourages fruit set.There are also varieties high in beta carotenes and vitamin A, hollow tomatoes and tomatoes which keep for months in storage.

Tomatoes are also commonly classified as determinate or indeterminate. Determinate, or bush, types bear a full crop all at once and top off at a specific height;

they are often good choices for container growing. Determinate types are preferred by commercial growers who wish to harvest a whole field at one time, or home growers interested in canning. Indeterminate varieties develop into vines that never top off and continue producing until killed by frost. They are preferred by home growers and local-market farmers who want ripe fruit throughout the season. As an intermediate form, there are plants sometimes known as "vigorous determinate" or "semi-determinate"; these top off like determinates but produce a second crop after the initial crop. The majority of heirloom tomatoes are indeterminate, although some determinate heirlooms exist.

Most modern tomato cultivars are smooth surfaced, but some older tomato cultivars and most modern beefsteaks often show pronounced ribbing, a feature that may have been common to virtually all pre-Columbian cultivars. While virtually all commercial tomato varieties are red, some tomato cultivars - especially heirlooms - produce fruit in colors other than red, including yellow, orange, pink, black, brown, ivory, white, and purple, though such fruit is not widely available in grocery stores, nor are their seedlings available in typical nurseries, but must be bought as seed, often via mail-order. Less common variations include fruit with stripes (Green Zebra), fuzzy skin on the fruit (Fuzzy Peach, Red Boar), multiple colors (Hillbilly, Burracker's Favorite, Lucky Cross), etc.

There is also a considerable gap between commercial and home-gardener cultivars; home cultivars are often bred for flavor to the exclusion of all other qualities, while commercial cultivars are bred for such factors as consistent size and shape, disease and pest resistance, and suitability for mechanized picking and shipping, as well as their ability to be picked before fully ripening. Tomatoes grow well with 7 hours of sunlight a day. A fertilizer with the ratio 5-10-10, often sold as tomato fertilizer or vegetable fertilizer can be used for extra growth and production, but manure or compost work well, too.

PLANT CHARACTERS

Growth Habit

Growth habit refers to the pattern of plant development and tomato plants can be classified into two major types:

a) Indeterminate

The terminal bud or the stem remains vegetative and will continue to produce leaves and stein from the growing tip. Flower clusters are formed after every three leaves. Plants can grow almost indefinitely and consequently indeterminates tend to be much taller than determinates.

b) Determinate

The growing point in the axil of the last-formed leaf transforms into a flower cluster, after which vegetative growth on that stein ceases. Flower clusters are formed about every two leaves. Semi-determinates have six or more clusters with two leaves between clusters. Strong determinates tend to be bushy and short compared to indeterminates. The height of semi-determinates is intermediate between determinates and indeterminates.

Seed Germination

Seed of most varieties germinate at an optimum temperature range of 26-32°. Under optimal conditions seedlings break through the soil surface about four days after sowing. Seed sowing depth should be about 1-2 cm.

Shoot Development

After emergence, the cotyledons expand rapidly and the first true leaves develop at the growing point. Leaves grow alternately (2/5 phyllotaxy). Most varieties develop 7-11 leaves below the first inflorescence. After the first cluster, indeterminate varieties typically develop three leaves between flower clusters and six or more flower cluster on a single branch. Determinate types usually show two leaves between flower clusters and up to 5 clusters per branch. Side shoots arise from each leaf axil that lead to new branches, leaves, and flower clusters.

Flowering

Tomato is a day neutral plant, meaning that day length does not affect the number of days to flowering. The first flower of the first cluster starts to open about 55-60 days after sowing. The second cluster flowers about one week after the first and so on for determinate varieties; duration from first flowering to completion of flowering is about 45-60 days for determinates. Time length between flowering of sequential flower clusters for indeterminates is 10-14 days.

Fruit Development

Fruit require 40-60 days from flowering to full ripening. Full fruit size will be achieved in 20-30 days after flowering.

CULTURAL PRACTICES

Seedling Production

About 250 g of seed is required to produce enough seedlings to plant one hectare. Seed germination should be greater than 80%. Seedlings grown in individual containers produce healthier and more vigorous seedlings compared to seedlings grown in flats or seedling beds. Seedlings grown in beds or flats suffer root damage when the plants are pulled for transplanting.

Seedling Container Method

— Fill a 10-cm-diameter pot or seedling tray with holes for individual plants with a well-drained medium such as 2:5:1 (sand, compost, burnt rice hull).

— Maintain the pots/trays off the ground in sheltered area such as a plastic covered rain shelter.

— Plant 2-3 seeds/pot and thin to one plant 2- 3 days after emergence of the first true leaves.

Seedbed Method

— Choose a well-drained area not recently cropped with tomato. A site where air circulation is good will be advantageous.

— In soils where soilborne pathogens are a problem, burning rice straw on the seedbed or soil fiunigation may be needed.

— Broadcast and incorporate fertilisers at the rate of 40 g ammonium sulfate, 50 g calcium superphosphate, 30 g potassium chloride, and 2 kg of compost for each 1 mn^2 of bed area.

— Prepare raised seedbeds at a width of 0.8 m and a height of 15 cm or higher to allow good drainage. Plant the seeds in rows 6 cm apart at a depth of 0.5 - 1.0 cm. Cover the bed surface with a thin layer of compost or ash before thinly mulching with rice straw and covering with mesh screen at 30 cm high. About 60 g of seeds are required per 250 m 2 seedbed.

— Care must be taken to prevent the soil from drying out and forming a crust on the soil surface. This crust often prevents seedling emergence, resulting in poor stand. Sprinkler irrigation of the seedbed is preferable to surface irrigation.

— Seedlings must be thinned within 2 or 3 days after the first true leaves appear.

Climate

Ideal growing conditions are needed to produce high quality seed. Tomatoes grow best in the dry season under day temperatures of 21-25°C and night temperatures of 15-20°C. Vines will struggle to set fruit if temperatures exceed 30°C. Humidity is an important factor. Humidity levels higher than 60% at the time of fruit maturity will increase disease problems and reduce seed yields. Seed production during the rainy season generally leads to low yields and poor seed quality.

Field Requirements

Avoid planting your tomato seed crop in a field previously planted with tomato, pepper, eggplant or other solanaceous crops. This will prevent the build-up of diseases and insects. Growing tomato after paddy rice reduces the incidence of diseases and nematodes. Optimum soil pH for tomato is 6.0 to 7.0; disorders such as blossom end rot are common if soil pH is lower than 5.5.

Land Preparation

Soil types. Tomato can be grown on soil textures ranging from sand to heavy clay but silty clay loam soils are preferable. Soil pH should range from 6.0-7.0. Tomato production after paddy rice is advantageous because of lower incidence of bacterial wilt and nematodes.

Beds. Shaping the land. into beds and growing tomato on top of the bed facilitates furrow irrigation of the crop and drainage after rainstorms. Beds can be prepared in many ways. At the AVRDC, beds are made with a. mechanical bed shaper and the beds are about 1 in wide and furrows (ditches) are 50 cm wide. Bed height varies, depending upon the season. In the dry season the bed height is 20 cm and increased to 35 cm in the wet season.

Mulches. Mulching is the use of inorganic materials such as thin plastic sheets, or organic materials such as rice straw to cover the soil surface. Mulches can modify soil temperatures, reduce fertiliser leaching, decrease moisture evaporation, and improve weed control. At AVRDC we use gray plastic mulch for tomato production. The plastic mulch is covered with rice straw in the hot summer to lower the temperature of the plastic.

Cultural practices

Healthy plants produce healthy seeds. Pests should be controlled. Nutrient and water management should be optimal to achieve good fruit and seed yield. For more information on growing tomato.

Selecting parents

Hybrid seed production involves the crossing of a female line to a male line. Either line can be the female or male parent, but normally the best seed yielder is selected as the female parent. Both parents should be pure, preferably being selfpollinated for more than six generations (this is called inbreeding). The inbred parents are selected for their desirable traits (e.g., high yields, disease resistance, fruit quality, earliness, etc.).

Ratio of male to female plants

It is important to have plenty of pollen available for making hybrid crosses. Since tomato vines bloom profusely, a ratio of one male for every four female plants is recommended.

Sowing dates

Seeds of male plants are sown three weeks earlier to ensure that pollen is available from the beginning of hybridisation.

Isolation

Each tomato flower has both male and female parts. Nearly all lines, including modern varieties, are selfpollinating. In other words, the pollen from each flower's stamens pollinate only the style of the same flower. Isolation of male and female parental lines is not necessary. Male plants can be planted as close as two meters from female plants. There are a few exceptions. Currant tomatoes (*L. pimpinellifolium*) and potato-leaf types of L. esculentum generally have styles that protrude outside their blossom. These lines are more attractive to insects and thus, more likely to be cross-pollinated. Netting or caging to exclude pollinating insects is required if any of these types are used as female lines. Also, do not collect seeds of double fruits since double flowers are more prone to insect pollination.

Plant location and spacings

Male lines are planted in a different location to facilitate operations and avoid shading from competing plants. Select a sunny spot to promote maximum production of flowers and pollen. Male and female lines are planted in double-row raised beds, with centers of beds spaced 150 cm apart. For female lines, plants are spaced 50 cm apart within the row. Male plants are spaced 40 cm apart to maximise flower production per hectare.

Staking

The female parent is staked. Staking facilitates the handling of plants during emasculation and pollination. Staking also keeps the ripening fruits above the ground and prevents rotting. Plants are trellised along with plants from the adjoining bed so that work operations are done on the raised bed rather than in the furrow.

Among male lines, only indeterminate types need to be staked. If male lines are staked, trellising can be done within beds or across adjoining beds.

Removing off-types

The male and female lines must be 100% pure. Know the plant habit, leaf type, and immature fruit characters (e.g., shape, size, and shoulder coloring) of each parent. Regularly inspect the plants.

Remove any off-type (usually inferior) or virusinfected plants before hybridisation procedures begin. Symptoms of viruses include yellow mottling of leaves; severe curling, cupping or other distortion of foliage; and stunting of plants.

Emasculation

Self-pollination cannot be allowed in hybrid seed production. The female flower must be pollinated by the pollen from the male line. To prevent self-pollination, remove the stamens from the flower buds of the female line before they shed their pollen). This process is called emasculation.

Emasculation begins about 55-65 days after sowing. Flower buds from the second cluster which will open in two to three days are chosen for emasculation. The petals will be slightly out of the flower bud but not opened, and the corolla color is slightly yellow or even paler. Flowers from the first cluster are removed. Sterilise the forceps, scissors and hands by dipping them in 95% alcohol before emasculation is started. If gloves are used, these should also be dipped in 95% alcohol to prevent pollen contamination.

Use sharp-pointed forceps to force open the selected buds. Then, split open the anther cone. Carefully pull the anther cone out of the bud, leaving the calyx, corolla and pistil. To help identify the hybrid fruits from selfed fruits at the time of harvest, cut the corolla and calyx (all or two sepals).

Pollen Collection

Collect flowers from the male parent to extract pollen. The best time for pollen collection is during the early morning before the pollen has been shed. Avoid pollen collection on rainy days. Remove the anther cones from the flowers and put them in suitable containers, such as glassine, cellophane, or paper bag.

Dry the anther cones by placing them 30 cm below a 100-watt lamp for 24 hours. The lamp creates a drying temperature of about 30°C. Pollen can also be sun-dried, but avoid drying at midday when temperature is very high. Put the dried anther cones in a plastic pan or cup. Cover the cup with a fine mesh screen (200-300 mesh) and then seal it with a similar tight-fitting cup, serving as a lid.

Shake the cup about 10-20 times so that the pollen is collected in the "lid" cup. Transfer the pollen into a small convenient-tohandle container for pollination.

Fresh pollen is best for good fruit-set. It can be kept for one day at moderate room temperature. When weather conditions are not suitable for pollination, dried or dehydrated pollen can be stored in a sealed container (capsule or vial) and kept in the freezer for about a month. Without freezing, the pollen can be kept in an ordinary refrigerator for two to three days without any significant loss in viability.

The pollen should be taken from the freezer or refrigerator and kept closed until the container warms to room temperature. This will prevent the pollen from getting wet due to condensation.

Pollination

Emasculated flowers are generally pollinated two days later. Try to avoid pollination on rainy days. The corolla of the emasculated flower turns bright yellow, signalling that the stigma is ready for pollination. Dip the stigma into the pool of pollen in the pollen container or pollinate by touching the stigma with the tip of the index finger dipped in the pollen pool.

Pollination is usually done three times weekly over a three to five week period. Successful pollinations are easily seen within one week by the enlargement of the fruit. After breeding operations are completed, any non-crossed flowers on the female plants are removed to lessen the chance of contamination from selfed seeds before harvest.

P AND K MANAGEMENT

Three factors determine the quantity of each inorganic fertiliser nutrient that must be applied to make up the difference between that required by tomato and that available from the soil and other unmanaged sources:

- the total quantity of uptake required to achieve the expected tomato yield;
- the portion of that total quantity that is supplied by the soil and other unmanaged sources; and
- the fraction of the fertiliser that will be recovered.

The total quantity of N required to achieve the expected total yield (marketable + non marketable fruit) can be estimated by multiplying the total yield in t/ha, by 2.4. P and K uptake is estimated by multiplying N uptake by 0.35 and 1.45, respectively. The estimates are in Kg P_2O_5 and Kg K_2O per hectare. Yield is strongly influenced by the physical environment as well as by management practices. Estimates of expected yield should be based on local experience and they should be realistic. If an overly ambitious estimate is accepted, fertiliser rates will be too high.

Fertiliser recovery is a function of nutrient availability, fertiliser rate, the form of fertiliser, and crop and fertiliser management practices. At optimum N rates, recoveries can range from about 40% when the soil is depleted of N and the crop is highly productive, to nearly 0% when the soil contains abundant N in available form and or the crop is unproductive. Corresponding recovery rates for P and K are 8 and 45%, respectively. As for N, recovery will approach 0% when the nutrients are abundantly available and or the crop need is small. Irrigation management and planting geometry are among the crop production practices that affect recovery. But more importantly, fertiliser placement practices (e.g., surface *versus* broadcasted-plowed-in *versus* banded) and timing affect the rate of nutrient recovery.

Chemical reagents for extracting available P and K from a soil sample are well established. When soils are tested to determine how much P and K are available, fertiliser recommendations are valid only if the soil tests have been calibrated for representative soil classes. Tests for available N are problematic. Soil fertility specialists familiar with local soils and production practices will be able to advise on methods of placement, sources of nutrients (including the fertiliser N, P and K equivalents of animal manures and composts), and rates of nutrient recovery.

In summary, if a reasonable estimate of the N, P and K requirements of a tomato crop can be established, then the appropriate soil fertility management strategies can be determined by a knowledge of soil properties and the processes that determine a soil's capacities to buffer nutrient removals and the fates of excess nutrients. A discussion of these matters is beyond the scope of this guideline. However, national soil fertility management specialists should have sufficient knowledge of nutrient supplying power for major groups of soil, the effects of past management on nutrient accumulations or depletions, the fractions of fertiliser that a crop will recover, and the effects of soil properties on recovery rates.

Transpling

Harden the seedlings by slightly reducing water and exposing them directly to sunlight about 6-9 days before transplanting. A good seedling should be about 3-4 weeks old (4 or 5-leaf stage) and vigorous, not thin and etiolated. Transplant in the late afternoon or on cloudy clays to minimise transplant shock. For transplanting, insert the seedling in a hole so the cotyledons appear above the soil surface. Older or etiolated seedling should be buried deeper. Press soil firmly around the root. Top irrigate lightly to settle the soil. After transplanting is completed, irrigate the field as soon as possible.

Irrigation

A water deficit during any crop stage will reduce yields. However, the crop is most sensitive to deficits during flowering, somewhat less sensitive during yield formation (fruit bulking) and immediately after transplanting, and least sensitive during vegetative growth. Because indeterminate flower and form fruit continuously after the first cluster flowers, these varieties remain very sensitive to deficits throughout post-vegetative growth. Relative to a reference evapotranspiration rate (ETo), the actual ET of an unstressed crop is greatest when the crop canopy is at full development (ET=1.1 x ETo) and least during the slow growth period after transplanting (ET=0.4 to 0.75 x ETo). A coefficient of 0.4 is appropriate for a field that is infrequently irrigated assuming that intervals between rain are long (10 or more days). A coefficient of 0.75 is for a field that is frequently irrigated, assuming that when intervals between rains are short (i.e., the soil surface moisture is maintained by a combination of irrigation and rain). If mulch is applied the coefficient will be smaller.

The total ET of an unstressed 105-day crop which is irrigated at an average interval of about one week (or receives well distributed rainfall) is a function of ETo:

	Average daily ETo, mm/day			
	3	4	5	6
	total ET, mm/crop			
ET	310	415	520	625

Irrigation should be scheduled to assure that water in the root zone is never depleted to a level that causes canopy stress (although mild stress in the early afternoon may be unavailable when ETo is very large). Root zones are shallow during the slow growth period after transplanting. During that period, therefore, fields should be irrigated frequently but at rates that just recharge the root zone. As the crop develops, the volume of the root zone increases. Therefore the irrigation rate can be greater and the intervals between applications lengthen accordingly.

Although the quantity of irrigation water is a function of unstressed ET, a large percentage of the water that reaches a field is lost by deep drainage, lateral seepage, and surface runoff. The quantity to apply is estimated as ZET/Ea where EET is daily ET totalled from the last date on which the root zone was recharged minus effective rainfall, and Ea is a coefficient between 0.4 and 0.7 that adjusts for irrigation efficiency. Ea varies with ir rigation practices and soil properties. Local irrigation management specialists will know the appropriate value of Ea as well as the method used to calculate effective rainfall. The specialists also will be able to provide seasonal estimates of ETo or the coefficients required to compute it from pan evaporation or other agricultural meteorological data.

Trellising /Staking

Staking or trellising is the use of bamboo, wood, metal poles, or other materials to support the plant and keep the fruit and foliage off the ground. Staking can increase fruit yield and size, reduces the proportion of unmarketable fruit, and facilitates chemical spraying, and harvesting. Indeterminate varieties should be staked in the dry and summer seasons to facilitate pruning, pinching, and other cultural practices. Determinate tomato varieties should be staked in the summer season in order to avoid fruit contact with wet soil which will cause fruit rots.. Many staking arrangements are possible.

Fixing

Plants should be fixed securely to the stake or string supports. Rice stray, plastic strips, horticultural fixing tape, or other materials can be used for fixing. Fixing should be done so fi-uit clusters are well-supported.

Pruning

Pruning is the selective removal of side shoots to limit plant growth. Indeterminate varieties should always be pruned. Pruning may force early fruit maturity, increase average fruit size and uniformity, improve fruit flavor and increase content of vitamins A and 'C content. Pruning improves air circulation in the canopy which may hinder development of foliar diseases, facilitates chemical spray applications and eases harvest Without pruning, indeterminates will set fruit poorly in the wet season because too much photosynthate is diverted into vegetative growth instead of the fruit. The degree of pruning varies according to the season. In Taiwan, farmers prune indeterminate varieties to one stem per plant, allowing just the main stern to bear clusters. This results in earlier maturing and larger fruit. However, in the summer two branches are maintained, the main branch, and second branch below the first fruit cluster. The reason for two branches in the summer is that the fruit require more vine cover or foliage to protect the fruit from the intense solar radiation.

Pinching

Pinching is the removal of extra flowers on the cluster and farmers may do this in order to increase fruit size. This is an advantage if markets pay a premium for larger fruit.

Hormone application.

High temperatures may drastically reduce tomato fruit-set and consequently, fruit yield. Application of tomatotone (4-Chlorophenoxyacetic acid or CPA) may increase tomato fruit-set under high temperatures.

Steps in Hormone Application

1. Prepare 1 % tomatotone solution by diluting 10 ml tomatotone concentrate per liter of water.
2. Pour the solution into a sprayer bottle and fit a cup over the nozzle large enough to fit over a flower cluster
3. Entire clusters are treated when 3-5 flowers have opened. Fit the cup over the cluster and apply one squirt. Do not treat clusters multiple times.
4. Apply hormone to flowers only, not foliage.
5. Subsequent clusters on the plant should be treated when 3-5 flowers have opened.

It is important that the grower apply hormone to each flower cluster only once. Multiple applications to the same cluster may lead to fruit deformities. In order to keep track of which flowers have been sprayed the grower might mix a dye in with the hormone. The dye will remain on the flower cluster and act as a visual aid for the grower.

Fruit Ripening

The number of hybrid fruits produced per plant depends on the fruit size of the maternal parent. As a rule of thumb, maintain the following: 30 fruits for large-fruited parent; 40 fruits for medium-fruited parent; and 50 or more fruits for small-fruited parent. Hybrid fruits are easily recognized by their cut sepals.

Remove the naturally-pollinated (non-hybrid) fruits, if any, from the female plants. This removal will prevent the accidental mixture of non-hybrid with hybrid fruits. Furthermore, non-hybrid fruits will steal nutrition away from the ripening hybrid fruits.

The process of fruit ripening from the "mature-green" stage to the "full ripe" stage involves changes in color, fruit composition, aroma, flavor, texture, all of which are regulated by plant hormones and modified by genetic and environmental factors. Changes include:

- Starch degradation and glucose/fructose formation
- Loss of fruit chlorophyll
- Increase in citrate and malate
- Lycopene and Beta-carotene synthesis

Fruit softening due to increase in polygalacturonase & soluble pectins

Production of flavor and aromatic compounds

- Breakdown of toxic alkaloid a-tomatine

Tomato is a climacteric fruit, meaning that ripening is accompanied by great surges in ethylene and respiration. The height of the climacteric happens at the pink-stage.

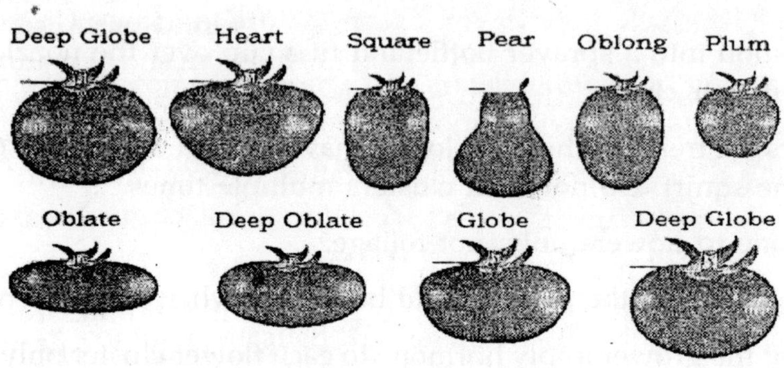

Figure 2. Different shapes of tomato fruits

Stages of fruit maturity

Fruit ripening is a continuous process but certain development stages are generally accepted:

- Immature Green: no jelly in locules
- *Green-mature:* jelly in locules; seeds not cut when fruit cut by shape knife. Fruit will mature normally if picked at this stage
- *Breaker:* not more than 10% of the surface is tannish-yellow, pink, or red; Color first appears on the blossom end
- *Turning:* color change has occurred over 10-30% of the fruit
- *Pink:* 30-60% of the fruit surface has shown pink/red
- *Light red:* >60% but <90% fruit surface is red/pink
- *Red:* >90 fruit surface is red (no pink)

Harvesting

Tomato can be harvested at a number of ripeness stages, depending upon distance to the market and intended purpose for the fruit. For long distance transport, fruit should be harvested at the green-mature stage, whereas fruit destined for local sale can be harvested at the pink or breaker stage. Tomato for processing is harvested at the full ripe stage.

Maturity means that the fruit have developed sufficiently so normal ripening will occur. Tomato fruit harvested at the "mature green" stage or beyond will ripen completely. Fruit harvested at the "immature' green stage will not fully ripen. It will be yellowish to pale red and have poor flavor. Green-mature tomatoes: (1) sometimes turn whitish green or develop cream-colored streaks at the blossom end at mature green stage (2) may have a fruit surface showing a waxy gloss and the skin is not readily scraped off with the fingernail (3) brown corky tissue on the stein scar (4) Well-formed jelly-like matrix in the locules and seeds are pushed aside rather than cut after fruit is sliced with a sharp knife Green-mature fruit need about 1-5 days to reach the breaker stage at 20 °C. Higher temperatures hasten fruit ripening.

Post-harvest loss. Tomato fruit is a living organ and it continues respiration after harvest; starch and sugar reserves are oxidised and heat is produced. Higher temperatures increase the respiration rate. In addition, fruit begin to lose water after harvest. Inadequate skill and care in fruit harvesting and handling, poor packaging for shipment, poor design of shipping containers, lack of good storage facilities, inadequate and untimely transportation, lack of a standard grading system, lack of an organised marketing system all can contribute to high postharvest losses or lower quality.

Tomato fruits ripen about 50-60 days after pollination, but may take longer if temperatures are cool. Keep the fruits on the vine until they are fully mature, preferably to the pink or red ripe stage. This enables the seed to develop normally and fully. If fruits are harvested at an earlier stage, place them in a covered, cool dry place for three or four days until they become red ripe.

Be sure to check for the clipped sepal before harvesting fruit. Collect fruits in nonmetallic containers, such as nylon net bags, plastic buckets, or crates. Metal containers may react with acids in the tomato juice and affect seed viability. Hence, they should not be used.

Seed Extraction

Option 1: Manual extraction

Harvest the ripe fruits and keep them in nylon bags. Crush the fruits by trampling with feet. Put the bags of crushed fruits into big plastic containers and ferment to separate the gel mass embedding the seeds. To hasten the fermentation process, put weights over the bags or keep the fruits submerged in the liquid fruit mass.

The time of fermentation depends upon the ambient room temperature. If temperature is above 25°C, one day of fermentation may be sufficient. If cooler, two days of fermentation may be needed. Fermentation for more than three days may spoil the seeds' quality. To wash the seeds, put them in an open plastic container. Then fill the container with water and stir the seeds to allow the pieces of flesh and skin sticking on the seeds to float.

Incline the container and gently remove the floating refuse, making sure that the seeds remain at the bottom. Repeat the washing several times, adding fresh water to the container every time until all the flesh and gel are completely removed, leaving clean seeds at the bottom.

Option 2: Mechanical extraction

Mechanical seed extraction is used by large-scale operations. Put ripe fruits into a mechanical seed extractor for crushing and separation of the seeds and gel from the pulp. Gather the seeds and gel mass in a suitable container such as plastic tub or bucket.

Instead of fermentation, treat the seed-gel mass with 0.7% hydrochloric acid (HCl) at a rate of seven milliliters of HCl per kilogram of seed-gel mass. Stir the seed-gel mass while the acid is being added. Continue stirring for 40 minutes until the gel is visibly softened or dissolved. Do not use a higher concentration of acid nor a longer treatment time, otherwise you will injure the seeds.

When the seed is separated from the gel, pour the acid-treated seeds into a clean fine-mesh bag. Wash the bag with tap water thoroughly so that no acid is left on the seeds' surface. While washing, step on the bag to squeeze out the remaining gel. Place the seeds into a plastic container, filling it to one-third capacity. Then, fill up the container with tap water. Stir the seeds to enable the small pieces of flesh and skin to float. Incline the container and remove the floating debris. Make sure the seeds remain at the bottom of the container.Repeat the washing procedure several times until all the debris is gone and the seeds are clean.

Precautions to Reduce Harvest Losses.

Harvest on time and pick selectively. Do not mix healthy and damaged fruit.

— *Harvest carefully.* Avoid physical injury to the fruit. Wounding and bruising increase water loss via wounds and damaged fruit respire more than healthy fruit.

— Avoid high temperatures. Fruit exposed to long periods of high temperatures will generate alcoholic and off-flavors. Harvest during cool periods such as late afternoon of early morning. Shade the fruit in the field and avoid temperatures >27 °C.

Avoid fruit dehydration. Maintaining fruit in a relative humidity range of 85-90% is ideal. RI-I below 80% will cause high fruit water loss.

— Pack fruit in containers that provide support and allow adequate ventilation to allow heat dissipation.
— Vehicles transporting the fruit to market should have good suspension to reduce jolting, especially on poor roads. Avoid over-loading.

Fruit grading. Standard grades vary among countries but grading is based on common criteria including fruit appearance i.e. size, color, shape, uniformity; condition and absence of defects.

Hybrid Tomato Production

Produce seed in cool, dry environments. The AVRDC produces hybrid seed in the dry season when temperatures generally do not exceed at mean of 28°C and the mean night temperature is 15-20°C

— Grow 4-5 female plants to 1 male plant
  — Sow the male parent about 7-10 days before the female parent so that plentiful pollen is available when the female parent is ready
  — Spacing for the female parent is 60 cm between rows in a two-row raised bed and 50 cm between hills. The female parent should be staked
  — Pollinate 40-50 flowers per female plant. As a rule of thumb, try to obtain 20 fruits per plant from large-fruited varieties, 30 fruits from medium-sized types, and >30 from small-fi uited types
— Plant the male parent spaced 150 cm between rows (single raised bed) and 40 cm between. hills. To promote good pollen production, grow the male in a sunny part of the field.

Pollen collection. It is best to collect pollen from freshly opened flowers in the morning. For small - it is easiest to collect pollen by use of a vibrator equipped. with a glass collection cup. The vibrator shakes the pollen from the anther cones and it is collected in the cup.

Pollination. Flowers emasculated 2 clays earlier should be ready for pollination. With a pair of scissors, cut the corolla and the calyx of the emasculated flower buds to expose the stigma and mark the hybrid pollinations. Dip the stigma into the pollen. Remove selfed flowers from the female or clearly indicate those flowers.

Typically, the pollination period will last about 30 days. The time from pollination until fruit harvest is about 45 days, depending upon the temperatures. Allow the fruit to ripen on the vine until the pink or red stage. Collect ripe fruit in a nylon mesh bag. Fermentation can be done in the same bag.

Seed extraction. Tomato seeds are enveloped in a mucilaginous sheath and fermentation hastens rotting of the gel. Harvest the fruits in nylon bags or another suitable container. Stomp on the fruit until fully crushed and the gel from the fruit is squished out. Gather bags in a large container for fermentation. If ambient temperatures are >25°C, then 1 day of fermentation is sufficient. If <25°C, then allow 2 days fermentation. HCL can be used for extraction of large seed quantities instead of fermentation. At the AVRDC, 0.7% HCL or 7ml HCL per kg wet seed is added for 20-30 minutes.

Seed washing. Loosen the net bag and pour the fermented pulp into *a* pail *so* that it occupies less than half the pail space, then fill the pail with water. Stir the mixture to break up caked portions of the -fermented pulp. The seeds will sink while the fermented remnants will float. Pour out the refuse but do not pour out the seeds. Add fresh water and repeat the procedure several times.

Seed drying. Place clean wet seeds into a nylon bag and dry using an electric dehydrator or clothes tumble dryer. Sun drying is also acceptable. Spread out the seed on an aluminium or plastic flat pan and enclose the container in a nylon bag. Dry the seeds at 28-30°C in an air drier for 2-3 days, making sure to turn the seeds several times a day. Moisture content of the dried seeds should be about 6%. Keep seeds in a cool dry place until packed into containers or packages. Seeds can remain viable for 10 years if stored at 5°C at 40% RH.

Seed yields: one ton of fruit produces about 10 kg seed.

Hybrid tomato varieties have many advantages compared to open-pollinated varieties. Hybrids usually produce higher yields. They generally mature earlier and more uniformly. Many hybrids have better fruit quality and disease resistance. With all of these advantages, many farmers prefer to sow hybrid seeds in spite of the higher seed costs.

The demand for hybrid tomato seeds can open a new market for growers interested in seed production. Hybrid tomato seed production is not easy. First, it requires much labor. Fortunately, this is not a problem in developing countries where affordable labor is available. Second, it requires the mastery of special skills and a close attention to detail. This publication will teach these skills.

TOMATO MARKET TYPES

There are three major market types in tomato based primarily on fruit type and intended use:

a) Fresh market

Fresh market tomato is used for cooking or eating raw. Most consumers prefer dark red tomatoes, globe or deep-glove shape, and 50-125 g size. However, fresh market tomatoes can vary greatly in color, shape, and size.

b) Cherry tomato

Cherry tomato fruit is small-sized (< 30g), and borne on long clusters (> 10 fruit/cluster). Cherry tomato is often sold in the fruit market rather than the vegetable market and good taste is particularly important.

C) Processing tomato

Processing tomato is produced for processing into paste, catsup, sauce, diced or - whole-peel tomato. Particular fruit quality such as dark red color, high solids (brix) content and pH <4.5. In addition to the fruit characters, processing varieties are usually of determinant growth type and some have compact vine and jointless pedicel to facilitate machine harvest.

FRUIT QUALITIES

Color

The red colored fruit of most tomato varieties is due to a carotenoid called lycopene. Lycopene synthesis is temperature sensitive and little is produced at temperatures >28 C. Tomato fruit ripened under high temperatures often develop a condition called "blotchy ripening" where lycopene does not form properly. Varieties grown in the off-season should develop adequate leaves (canopy or vine cover) to shade the fruit from the sun.

Size

Depending upon the variety, average fruit range can range in size from 5->300 grams. Fruit borne on lower clusters tend to be larger than those on the upper clusters; on a given cluster, proximal fruit (those closest to the stein) tend to be larger than distal fruit. Production of large-fruited varieties is difficult under high temperatures and. heat tolerant varieties tend to be small-fruited. At the AVRDC we have found that fruit size of heat tolerant varieties is reduced by 30% or more when grown in the summer compared to the dry season.

Shape

Fruit shape ranges from almost flat to elongated types. In most countries, consumers prefer globe, deep globe, oblong, or pear shapes.

Firmness

Fruit firmness can be defined as the resistance of the fruit pericarp to pressure and thick walls are often associated with firmness. Finnness is a critical quality character because it determines whether or not *fruit can* be shipped to distant markets. Tomato fruit soften as they develop from immature green to full red color.

Solids (°brix)

Tomato fruit is about 95% water, and 4-5% organic compounds called solids. The solids portion consists of about 50% sugar (glucose and fructose) found mostly in the fruit wall; 25% is alcohol insoluble solids which. include pectins, cellulose, proteins, polysaccharides; and organic acids, mostly citrate and malate. The remainder of the solids consists of carotenoids, volatile compounds, amino acids, and inorganic compounds.

A high solids content is important for processed tomato, especially paste. However, yield and solids content are negatively correlated (the higher the yield the lower the solids). In general, indeterminate varieties tend to have higher fruit solids contents compared to determinate types.

Acidity ((Xi citrate)

Most acid in the fruit is contained in the locules and acidity ranges in pH from 4-5 in tomato. A pH <4.5 is required for processed tomato because microbial growth is inhibited. A high acid content imparts a sour taste that is desirable in some countries.

REFERENCES

Food and Agriculture Organization of the United Nations. 1986.*Improvement of post-harvest fresh fruits and vegetables handling.* FAO,Bangkok, Thailand.

Gould, W.A. 1992. *Tomato production, processing, and technology.* ThirdEd. CTI Publications, Inc, Baltimore.

Grierson, D., and A.A. Kader. 1986. Fruit ripening and Quality. Pp. 241-280. In: *The Tomato Crop* (J.G. Atherton and J. Rudich, eds.) Chapman and Hall,London.

Kader, A.A., and L. Morris. 1976. Appearance factors other than color andtheir contribution to quality. Pp. 8-14. In: Proceedings of the Second TomatoQuality Workshop. University of California, Davis, USA

Ryall, A.L., and W.J. Lipton. 1979. *Handling, Transportation, and Storageof Fruits and Vegetables.* AVI Publishing, Westport, Conn.

Villareal, R.L. 1980. *Tomatoes in the Tropics.* Westview Press,Boulder, Colorado.

16

Mushroom

Mushrooms are the fruiting bodies of macrofungi. They include both edible/medicinal and poisonous species. However, originally, the word "mushroom" was used for the edible members of macrofungi and "toadstools" for poisonous ones of the "gill" macrofungi. Scientifically the term "toadstool" has no meaning at all and it has been proposed that the term is dropped altogether in order to avoid confusion and the terms edible, medicinal and poisonous mushrooms are used.

Edible mushrooms once called the "food of the gods" and still treated as a garnish or delicacy can be taken regularly as part of the human diet or be treated as healthy food or as functional food. The extractable products from medicinal mushrooms, designed to supplement the human diet not as regular food, but as the enhancement of health and fitness, can be classified into the category of dietary supplements/mushroom nutriceuticals. Dietary supplements are ingredients extracted from foods, herbs, mushrooms and other plants that are taken without further modification for their presumed health-enhancing benefits.

There is an old Chinese saying which states that "Medicines and foods have a common origin". Mushrooms constitute a most rapidly growing new food category which the current health-oriented public is increasingly enjoying.

Since mushrooms lack chlorophyll they can not, like green plants, get their energy from the sun through photosynthesis. Instead, during their vegetative growth stage, mushroom mycelia secrete enzymes that break down compounds such as cellulose and lignin present in the substrate. The degraded compounds are then absorbed by the hyphae and the mycelium enlarges-usually laterally, and in some cases growing several meters in diameter with the substrate.

Partially understood environmental factors (temperature and light are known to be critical) stimulate the second or reproductive growth stage. Cells of one mycelial strain fuse with cells of the opposite type to form a mycelium that contains both types of nuclei.

The new mycelium continues to grow and eventually develops into a mature fruiting body, the gills of which are lined with spore bearing cells called basidia. Various mechanisms trigger the dispersal of spores, which in turn lodge in a substrate, become hyphae and begin the cycle anew.

Mushroom cultivation has great scope in China, India and in some of other developing countries because of the cheap and easily available raw materials needed for this activity, coupled with faster means of communication and marketing (as a fresh commodity), and better purchasing power of the people. Using China as for example, in 1978, the production of edible mushrooms was only 60,000 tonnes. In 2006, China's mushroom production was over 14 million tonnes. Now there are more than 30 million people directly or indirectly engaged in mushroom production and businesses, and now China has become a leading mushroom producer and consumer in the world.

It is hoped that the avocation of mushroom farming will become a very important cottage industry activity in the integrated rural development programme, which will lead to the economic betterment of not only small farmers but also of landless labourers and other weak sections of communities. The advantages of mushroom cultivation can be summarised as:

1. Wastes such as cereal straws are largely burnt by the farmers, which causes air pollution. However, these raw materials can actually be used for the cultivation of mushrooms. This kind of bioconversion exercise can greatly reduce environmental pollution.
2. Mushroom cultivation can be a labour intensive activity. Therefore, it will serve as means of generating employment, particularly for rural women and youths in order to raise their social status. It will also provide additional work for the farmers during winter months when the farming schedule is light.
3. It will provide the people with an additional vegetable of high quality, and enrich the diet with high quality proteins, minerals and vitamins which can be of direct benefit to the human health and fitness. The extractable bioactive compounds from medicinal mushrooms would enhance human's immune systems and improve their quality of life.
4. Mushroom cultivation is a cash crop. The harvested fruiting bodies can be sold in local markets for additional family income or exported for an important source of foreign exchange that will definitely improve the economic standards of the people.
5. Some warm mushrooms, e.g. Volvariella volvacea (Straw mushrooms) and Pleurotus sajor-caju (Oyster mushrooms) are relatively fast growing organisms and can be harvested in 3 to 4 weeks after spawning. It is a short return agricultural business and can be of immediate benefit to the community.

Mushroom farming is both a science and an art. The science is developed through research the art is perfected through curiosity and practical experience. However, mushroom farming is a business which requires precision. Indeed, it is not as simple as what some people often loosely stipulate. It calls for adherence to precise procedures. If you ignore one critical step, you are inviting trouble, which could lead to a substantially reduced mushroom crop yield. For example, if you fail to adjust the pH of the substrate to a critical level required by the specific mushroom species your are cultivating, or if you do not properly pasteurise the substrate, your planted mushroom could be outcompeted by unwanted, intrusive micro-organisms.

The following ideas need to be emphasized as a conclusion of this introduction. Although scientific research and farming practice have led to the development of some universal or general concepts concerning mushroom cultivation, the diverse biological nature of the process also means that a wide spectrum of variations in farming methods must be employed. Thus the transfer of mushroom farming from one region or country to another cannot be treated in the same way as the transfer of non-biological industrial technology, such as that of complete complex of factory equipment for textile or chemical fertiliser industries. Since the cultivation of mushrooms deals with living organisms, one should consider, not only the unique attributes of the mushroom itself, and of the various micro-organisms growing with it, but also the physical and biochemical natures of the substrates. Therefore, the cultivation methods must be tailored in accordance with the prevailing unique natural resources, heritage, local climate, and socio-economic conditions of each farming community. All these considerations call for a critical mass of well trained mushroom scientists and mushroom growers. Thus the training activities need to be supported by UN agents as well as Governments of the various countries concerned.

MUSHROOM BIOLOGY

It has been well known that the 20th century has been an explosive time for the accumulation of knowledge. Modern technology for human civilisation is expanding every day. However, human beings still face and will continue to face three basic problems: shortage of food; pollution of the environment; and diminishing quality of human health, due to the continued increase of the world population. The 20th century began with a world populated by 1.6 billion people and ended with 6 billion inhabitants—with most of the growth occurring in the developing countries. The growing world population is increasing by about 80 million people per year. At the present, about 800 million people in the world are living in poverty. On the other hand, it has been observed that over 70 % of agricultural and of forest products has not been put to total productivity, and have been wasted in processing. Macrofungi (mushrooms) not only can convert these huge lignocellulosic biomass wastes into human food, but also can produce notable immune enhanced products, which have many health benefits. Another significant aspect of mushroom cultivation is using the biota in creating a pollution-free environment.

Mushrooms with other fungi are something special in the living world, being neither plants nor animals. They have been placed in a kingdom of their own called the kingdom of Myceteae. But what are mushrooms? The word mushroom may mean different things to different people and countries. It has emerged that specialised studies and the economic value of mushrooms and their products had reached a point where a clear definition of the term "mushroom" was warranted. In a broad sense "Mushroom is a macrofungus with a distinctive fruiting body, which can be either epigeous or hypogeous and large enough to be seen with naked eye and to be picked by hand". Thus, mushrooms need not be basidiomycetes, nor aerial, nor fleshy, nor edible. Mushrooms can be ascomycetes, grow underground, have a non-fleshy texture and need not be edible. This definition is not a perfect one but can be accepted as a workable term to estimate the number of mushrooms on the earth. The most common type of mushrooms is umbrella shaped with a pileus (cap) and a stipe (stem) i.e. *Lentinula edodes*. Other species additionally have a volva (cup) i.e. *Volvariella volvacea* or an annulus (ring) i.e. Agarius campestris or with both of them i.e. *Amanita muscaria*. Furthermore, some mushrooms are in the form of pliable cups; others round like golf balls. Some are in the shape of small clubs; some resemble coral; others are yellow or orange jelly-like globs; and some even very much resembles the human ear. In fact, there is a countless variety of forms.

The structure that we call a mushroom is in reality only the fruiting body of the fungus. The vegetative part of the fungus, called the mycelium, comprises a system of branching threads and cord-like strands that branch out through soil, compost, wood log or other lignocellulosic material on which the fungus may be growing. After a period of growth and under favourable conditions, the established (matured) mycelium could produce the fruit structure which we call the mushroom. Accordingly mushrooms can be grouped into four categories: (1) those which are fleshy and edible fall into the edible mushroom category, e.g., *Agaricus bisporus*; (2) mushrooms which are considered to have medicinal applications, are referred to as medicinal mushrooms, e.g., *Ganoderma lucidum*; (3) those which are proven to be, or suspected of being poisonous are named as poisonous mushrooms, e.g., *Amanita phalloides*; and (4) a miscellaneous category which includes a large number of mushrooms whose properties remain less well defined, which may tentatively be grouped together as 'other mushrooms'. Certainly, this approach of classifying of mushrooms is not absolute and not mutually exclusive. Many kinds of mushrooms are not only edible, but also possess tonic and medicinal qualities.

Mushrooms are devoid of leaves, and of chlorophyll-containing tissues. This renders them incapable of photosynthetic food production. Yet, they grow, and they produce new biomass. How? For their survival, for their growth, and for their metabolism, they rely on organic matter synthesized by the green plants around us, including organic products contained in agricultural crop residues. The organic materials, on which mushrooms derive their nutrition, are referred to as substrates. Mushrooms are a unique biota which assembles their food by secreting degrading enzymes and decompose the complex food materials present in the biomass where they grow, to generate simpler compounds, which they then absorb, and transform into their own peculiar tissues. These substrate materials are usually by-products from industry, households and agriculture and are usually considered as wastes. And these wastes, if carelessly disposed of in the surrounding environment by dumping or burning, will lead to environmental pollution and consequently cause health hazards. However, they are actually resources in the wrong place at a particular time and mushroom cultivation can harness this waste/resource for its own beneficial advantage.

Mushrooms lack true roots. How then are they anchored into the substrates where we find them? This is affected by their tightly interwoven thread-like hyphae, which also colonise the substrates, degrade their biochemical components, and siphon away the hydrolysed organic compounds for their own nutrition.

Mushroom Hunting

Fungi are found just about everywhere. Mushrooms, a special group of macro-fungi, are rather more selective than other fungi in that the size of the fruiting body requires the availability of more nutrients than are required for the production of asexual spores by micro-fungi. In damp places, such as tree-fern ecosystems and tropical rain forests, plentiful moisture leads to abundant mushroom formation. There, mushrooms can be collected at most times of the year. But in drier regions, they occur only after seasonal rains. In these ecosystems there may be a particular flora of mushroom species associated with the seasons of autumn, summer and spring. Relatively few mushrooms are produced during the cold winter months, although there are perennial fruiting bodies that persist during the winter. The formation of mushroom fruiting bodies depends very much on the pattern of rains and, in some years, there may be virtually total lack of mushroom fruiting.

Mushroom hunters, in addition to carrying along with them the basic equipment and field guide references, which will vary depending on personal requirements and regional conditions, should record such items as date, time, location, smell, substrate (host) colour, habitat and anything at all unusual about the specimen. Some important characteristics for identification disappear rapidly as the mushroom matures. These characteristics have to be recorded accurately at the time of collection.

Ecological Classification

Ecologically, mushrooms can be classified into three groups: the saprophytes, the parasites and the symbiotic (which include mycorrhizal) species (Fig. 1). There are only a few parasitic mushrooms. Most of the cultivated gourmet mushrooms are saprophytic fungi. Some of the edible mushrooms are mycorrhizal species, e.g. Perigold black truffle, Tuber melanosporum, and matsutake mushroom, Tricholoma matsutake. It is difficult to bring these highly celebrated wild gourmet species into cultivation because they are mycorrhizal. These species have a symbiotic relationship with some vegetation, particularly trees, i.e. there is a relationship of mutual need. Saprophytes obtain nutrients from dead organic materials; parasites derive food substances from living plants and animals and causing harm to the hosts; and mycorrhiza live in a close physiological association with host plants and animals – thereby forming a special partnership where each partner enjoys some vital benefits from the other. However, some mushrooms do not fall neatly within these man-made categories and can share two of these categories. For example, some Ganoderma spp. including G. lucidum are common saprophytes, however they can be pathogenic too; also Tricholoma matsutake, while initially appearing to be mycorrhizal on young roots, soon becomes pathogenic and finally exhibits some saprobic ability.

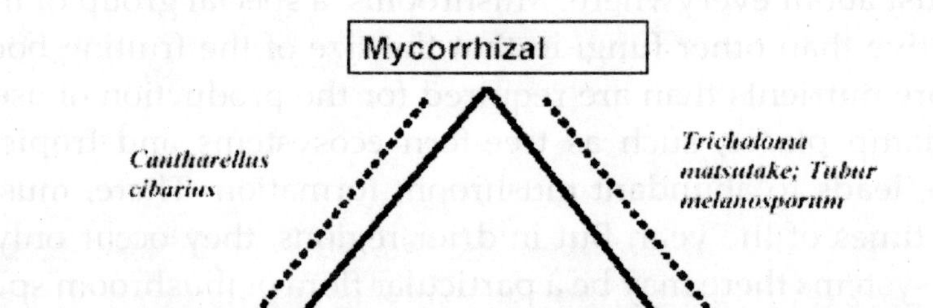

Figure 1. Modified triangular model for the ecological classification of mushrooms.

Magnitude of Mushroom Species

In 1990, the number of known species of fungi was about 69,000 while it was conservatively estimated that 1.5 million species of fungi actually existed. On average 700 species were described as new to science each year from 1920 to 1950. However, the annual total catalogued fungi reached around 1,400 in 1961, 1,500 in 1968 and averaged 1,700 each year for 1986 to 1990.

Fungi are regarded as being the second largest group of organisms in the biosphere after the insects. Known fungal species constitute only about 5% of their species in the world. Thus, the large majority of fungi are still unknown. Out of about 70,000 described species of fungi, it has been suggested that around 14,000-15,000 species produce fruiting bodies of sufficient size and suitable structure to be considered as macrofungi (mushrooms). Of these, about 5,000 of the species are considered to possess varying degrees of edibility, and more than 2,000 species from 31 genera are regarded as prime edible mushrooms. But only 100 of them are experimentally grown, 50 economically cultivated, around 30 commercially cultivated, and only about 6 to have reached an industrial scale of production in many countries. Furthermore, about 1,800 are medicinal ones. The number of poisonous mushrooms is relatively small (approximately 10%), of these some 30 species are considered to be lethal.

Table 1. Comparison of the numbers of known and estimated total species in the world of selected groups of organisms

Group	*Known species*	*Total species*	*% known species*
Vascular plants	220,000	270,000	81
Bryophytes	17,000	25,000	68
Algae	40,000	60,000	67
Fungi	69,000	1,500,000	5
Bacteria	3,000	30,000	10
Viruses	5,000	130,000	4

The biological science that is concerned with fungi is called mycology. Mushroom biology is the branch of mycology that deals with mushrooms in many disciplines. When knowledge increases and areas of specialisation develop within the discipline, it is convenient to indicate that area of specialisation with a self-explanatory name. In biology, there are such specialisations as neurobiology, bacteriology, plant pathology, pomology, molecular biology, virology, fungal physiology, embryology, endocrinology, phycology, and entomology. These names indicate either a group of organisms (e.g., bacteria, algae, and insects) and /or an approach to the study (e.g., disease, development and physiology).

Although several terms for this important branch of mycology that deals with mushroom have been used, and each of these has its merit, when we get down to the matter of definitions, it seems that there is a place for a new term. The new term is mushroom biology. Mushroom biology is a new discipline concerned with any aspect of the scientific study of mushrooms, such as: taxonomy; physiology; genetics; etc.

Applied Mushroom Biology

Applied mushroom biology is concerned with all aspects of the application of mushroom biology. It consists of three main components: mushroom science; mushroom biotechnology; and mushroom mycorestoration (Fig. 2). As previously outlined, mushroom biology is concerned with any aspect of the scientific study of mushrooms, therefore it will feature in each of the three components of applied mushroom biology.

Mushroom science deals with mushroom cultivation and production (mushrooms themselves) and encompasses the principles of mushroom biology/microbiology, bioconversion/composting technology and environmental technology.

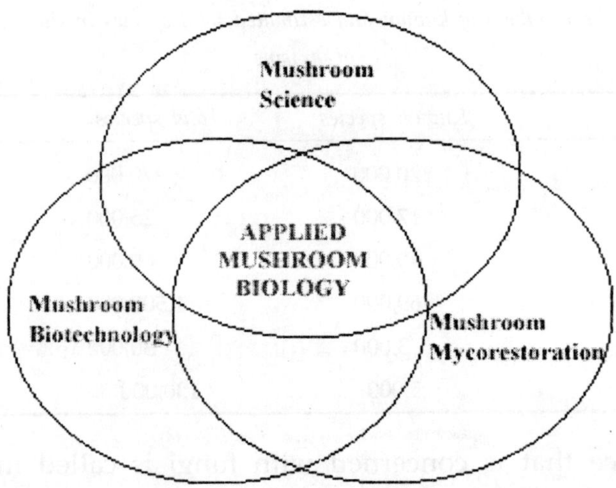

Figure2. The three components of Applied Mushroom Biology: Mushroom Science; Mushroom Biotechnology; and Mushroom Mycorestoration.

Mushroom biotechnology is concerned with mushroom products (mushroom derivatives) and encompasses the principles of mushroom biology/microbiology, fermentation technology and bioprocess. Mushroom biotechnology, both as a technology and as the basis for new mushroom products, requires industrial development. It, like many bioscience industries, operates at the cutting edge of science and involves numerous regulatory issues.

The third component of applied mushroom biology has been developed in recent years. This is mushroom bioremediation which is concerned with the beneficial impacts of mushrooms on the environment (from mushroom mycelia) and encompasses principles of mushroom biology/microbiology, ecology and bioremediation technology.

Therefore, the aims of the discipline of applied mushroom biology are to tackle the three basic problems: shortage of food, diminishing quality of human health and pollution of the environment, which human beings still face, and will continue to face, due to the continued increase of the world population. The 20th century began with a world population of 1.6 billion and ended with 6 billion inhabitants. The world's population is likely to reach 9.2 billion in 2050 from the current 6.7 billion with most of the growth occurring in developing countries. The growing world population is increasing by about 80 million people per year. At present, about 800 million people in the world are living in poverty. On the other hand, it has been observed that over 70% of agricultural and of forest products have not been put to total productivity, and have been discarded as waste. Applied mushroom biology not only can convert these huge lignocellulosic biomass wastes into human food, but also can produce notable nutriceutical products, which have many health benefits.

Food Supplies through Mushroom

The discipline that is concerned with the principles and practice of mushroom cultivation is known as mushroom science. The establishment of principles requires facts, which are arrived at through systematic investigation. The systematic investigation must involve the practical aspects of mushroom cultivation as well as scientific studies. The consistent production of successful mushroom crops necessitates both practical experience and scientific knowledge.

Mushroom Biotechnology

It has been pointed out that mushroom biotechnology is concerned with mushroom products and encompasses the principles of fermentation technology, mushroom biology/microbiology and bioprocess. Mushroom products have a generalised or tonic effect, which in some cases may act prophylactically by increasing resistance to disease in humans from the balancing of nutrients in the diet and the enhancing of the immune systems.

Nutritional value of mushrooms

The greatest difficulty in feeding man is to supply a sufficient quantity of the body-building material—protein. The other three nutritional categories are: the source of energy food—carbohydrates and fats; accessory food factors—vitamins; and inorganic compounds which are indispensable to good health. Of course, water, too, is essential. In terms of the amount of crude protein, mushrooms rank below animal meats, but well above most other foods, including milk, which is an animal product. Furthermore, mushroom protein contains all the nine essential amino acids required by man. The moisture content of fresh mushrooms varies within the range 70 - 95% depending upon the harvest time and environmental conditions, whereas it is about 10 - 13% in dried mushrooms. In addition to their good proteins, mushrooms are a relatively good source of the following individual nutrients: fat, phosphorus, iron, and vitamins including thiamine, riboflavin, ascorbic acid, ergosterine and niacin. They are low in calories, carbohydrates and calcium. Mushrooms also contain a high proportion of unsaturated fat. In recent years, there has been a trend toward discovering ways of treating mushrooms so as to give them added value. For example, Wermer and Beelman have reported on growing mushrooms enriched in selenium. The desirability of a food product does not necessarily bear any correlation to its nutritional value. Instead, its appearance, taste, and aroma, sometimes can stimulate one's appetite (preference). In addition to nutritional value, mushrooms have some unique colour, taste, aroma, and texture characteristics, which attract their consumption by humans.

Medicinal value of mushrooms

The second major attribute of mushrooms, their medicinal properties, has also been drawn to our attention for study, e.g., hypotensive and rental effects, immunomodulatory and antitumour activities of polysaccharide-protein complex (PSPC) from mycelial cultures, immunomodulatory and antitumour activities of lectins from edible mushrooms, isolation and characterisation of a Type I Ribosome-Inactivation protein from V. volvacea, and medicinal effects of Ganoderma lucidum.

Nutriceuticals and dietary supplements

The recent upsurge of interest in traditional remedies for various physiological disorders and the recognition of numerous biological response modifiers in mushrooms have led to the coining of the term "mushroom nutriceuticals". A mushroom nutriceutical is a refined/partially defined mushroom extractive which is consumed in the form of capsules or tablets as a dietary supplement and which has potential therapeutic applications. A regular intake may enhance the immune responses of the human body, thereby increasing resistance to disease and, in some cases, cause regression of a disease state.

Mushroom bioremediation: Benefit the Environment through Mushroom Mycelia This component of applied mushroom biology deals mainly with the aspects of benefits to the Earth from the activities of mushroom mycelium. Environmental contamination can be ameliorated by the application of mushroom mycelial technologies. For example, (1) the use of bioconversion processes to transform the polluting substances into valuable foodstuffs, e.g., the proper treatment and reutilisation of spent substrates/composts in order to eliminate pollution problems. One of the most intriguing opportunities offered by mushroom mycelia in the area of bioconversion is the exploitation of their ability to degrade pollutants, many of which are highly carcinogenic, released into the environment as a consequence of human activity. And (2) the use of fungi/mushroom mycelia as tools for healing soil, what Stamets called "mycorestoration", which is the use of fungi/mushrooms to repair or restore the weakened or damaged biosystems of environment. The processes of mycorestoration include the selective use of mushrooms for mycofiltration, to filter water; mycoforestry, to enact ecoforestry policy; mycoremediation, to denature toxic wastes; and mycopesticides, to control insect pests. Mycoresoration recognises the primary role fungi/mushrooms can play in determining the balance of biological populations.

Mushroom Genetics and Breeding

Studies of the genetics of fungi have played a leading role in the development of the modern era of molecular genetics. The chemical studies of Beadle and Tatum beginning in the 1940's on the biochemical genetics of the ascomycete Neurospora brought forth techniques that were subsequently employed with bacteria. Those studies then raised questions concerning the nature of the genetic material which culminated in the elucidation of the structure of DNA by Watson and Crick in 1953. Since that time, fundamental investigations on the molecular level have progressed with ever-increasing rapidity.

The basidiomycetes were not in the fore-front of research during the early days when molecular biologists preferred the rapidity of the cell cycle of the unicellular bacterium Escherichia coli to the slower growth of the filamentous basidiomycetes with their complex incompatibility control systems of sexuality. In recent years, however, basidiomycetes have attracted the interest of investigators concerned with problems of differentiation and the control of development because they are eukaryotic organisms and do have some distinct but simple morphological stages which are not present in single cell organisms. However, mushroom genetics is a young discipline in science as compared to biotechnology, which is correlated with the beginning of human civilisation. It was almost exclusively devoted to fundamental research before attracting serious attention during the early 1770s in relation to mushroom production. The overall objective of breeding is to improve the quality of extant strains in the most efficient way possible through the target selection, and controlled crossing, and progeny selection. The desired result should be defined with reasonable clarity in terms of measurable traits desired, such as yield, quality of flavour, texture, appearance of mushrooms, disease resistance, and general vigour, all within the context of a chosen set of standard conditions which are employed consistently throughout the breeding programme.

Discovery of sexuality by kniep and bensaude

Although Blakeslee in 1904 had demonstrated sexuality in the Mucorales through matings of mycelia established from single spores, it was not until 1918-20 that the mechanism of sexuality in the basidiomycetes was discovered. It was demonstrated that the mycelium arising from single spores was made up of hyphae with simple septa, and that the hyphae of the fruiting bodies and of mycelium giving rise to fruiting bodies bore clamp connections. Furthermore, when mycelia which had originated from single spores were confronted with various combinations, only certain combinations gave rise to clamped mycelium, while in other combinations the hyphae had simple septa.

Sexuality in the edible mushrooms

Although the process of sexuality is complicated by nutritional and physiological conditions, genetic constitution is the most critical factor determining both the occurrence and the morphology of the fruiting bodies in the edible mushrooms. Sexuality in fungi consists of three important stages. The first essential stage is plasmogamy which is the fusion of cytoplasm of the two mating individuals. By plasmogamy the nuclei from two strains are brought together in a common cytoplasm. The second essential stage in sexuality is known as karyogamy or nuclear fusion. The third essential stage is meiosis, the nuclear division in which the chromosome number is reduced from the diploid to the haploid number. The product of meiosis is the formation of a tetrad. Through the process of sexuality, genetic recombination and segregation subsequently occurs.

Mating Systems (Patterns of Sexuality) in Fungi

Edible mushrooms contain both self-fertile and self-sterile species. Self-fertilisation (homothallism or homomixis) is probably the commonest mode of sexual reproduction in the fungi as a whole but in the higher fungi (e.g., basidiomycetes) self-fertile species are in a distinct minority of only about 10 %. Self-sterilisation (heterothallism or dimixis) is about 90 %, in which 25 % is bipolar and 65 % is tetrapolar.

1) *Homothallism.* Fruiting structure can be produced by a single, monosporous mycelium. Potentially self-fertile fungi are not always necessarily homozygous and a variety of situation and rather imprecise regulating system can result in heterozygosity. Two types of homothallism are found among self-fertile species: (a) primary homothallism, in which a homokaryotic mycelium, established from a single meiotic nucleus, has the potentiality to progress through heterokaryosis to the completion of the sexual cycle; and (b) secondary homothallism, in which a fertile

dikaryotic mycelium is established from a basidiospore carrying two meiotic nuclei of different mating types.

2) Heterothallism. Cross mating between homokaryotic mycelia is required to complete the sexual cycle. Two mating systems of heterothallism are commonly found in edible mushrooms:

   (a) Bipolar mating system, in which the mating competence is determined by incompatibility factors of a single series, the A factor. Therefore, only two mating types are produced in equal frequency by a single fruiting body. (b) Tetrapolar mating system, in which the mating competence is determined by incompatibility factors of two series, A and B, which assort and segregate independently at meiosis. There are four rather than two mating types produced in equal frequency by a single fruiting body.

Life Cycle

If a section of the gills is cut and examined under the microscope, spores will be observed on their surface. The spores will start to fall as the cap fully expands, indicating maturity of the mushroom. The spores are so minute that they float in the air and are carried by the wind. Eventually, they fall to the ground, usually with rain. If conditions are favourable (optimum temperature and moisture), the spores will germinate to form a mass of mycelium. This is the start of the vegetative phase of the mushroom. Given an unrestricted amount of nutrients and favourable growing conditions, it is capable of unlimited growth. The mycelium developing from the germinating spore is the so-called primary mycelium and is usually uninucleate and haploid. This stage is short-lived because mycelia from different spores tend to ramify and fuse to form the secondary mycelium with two compatible nuclei, which continues to grow vegetatively and is able to form fruiting bodies.

Strain Improvement (Breeding Programme)

The reservoir of edible mushrooms, like other micro-organisms of use for industrial purposes, is not unlimited. It is generally recognised that in order to maintain and breed high-yielding strains, the techniques employed in mushroom breeding should now and then be modified and improved in accordance with new findings and progress in the scientific world as a whole, and in microbiology and genetics in particular.

1) *By selection.* The common white mushroom, Agaricus bisporus, strains are usually based on selection from multi-spore or single-spore cultures. Tissue culture of selected sporophores has also been used to fix desired variants. In the short term, selection alone appears to have some role to play in mushroom strain improvement,

but genetic improvement by selection becomes increasingly difficult. Therefore, strain improvement through hybridisation has become a recourse and also necessity.

2) *By hybridisation.* In addition to the conventional method of matings between two genetically compatible strains through which dikaryon mycelia and fruiting bodies are formed, steps towards a broader spectrum of hybridisation can be achieved in strain improvement of edible mushrooms by the following ways.

   a) *Use of auxotrophs.* Auxotrophs can be obtained naturally or induced by mutagenesis. The contrasting auxotrophs can be paired and the products can be screened for hybridity on minimal medium. Certainly, the feasibility of auxotrophs to be used as a tool for hybridisation depends on how easily auxotrophs can be obtained in the strains of the mushroom.

   b) *Use of resistance markers.* Mutants resistant to antimetabolites have been suggested recently as alternatives to auxotrophs for use in mushroom breeding programmes. The treated spores or hyphal fragments, which can grow on a medium containing an inhibitory concentration of the anti-metabolite, would be considered to possess the marker. Complementary resistant strains would be grown together and, hybridity can be confirmed by transferring it to a medium containing the two appropriate antimetabolites.

   c) *Protoplast fusion.* One of the most effective barriers to sexual reproduction is the inability of hyphae from two selected strains to fuse. Several laboratories have reported that protoplasts can be isolated from plant and microbial cells by enzymatic breaking of the cell wall, in the presence of an osmotic stabiliser. Such protoplasts can be effectively induced to fuse in the presence of polyethylene glycol (PEG). After a short period of time protoplasts can regenerate their cell walls and start to propagate as normal cells or hyphae. These cells are heterokaryotic if fusion occurred between cells from genetically different strains. This can serve to increase the frequency of intraspecies crosses in organisms in which natural matings rarely occur. The technique has even a much wider application, and can be used for interspecies and intergeneric crosses in some organisms, which normally cannot be crossed. Although such an approach has been carried out in several laboratories, until now, no clear and economically applicable results have been reported in edible mushrooms. I hope that this does not embrace the truth that "beautiful hypotheses are destroyed by ugly facts".

Due to the nature and the limited time of the training course, the molecular biology of mushrooms was not touched upon in this course, although it has become an important sector of mushroom genetics and breeding. For further reading, readers are referred to Chang et al. It should be emphasised that sexuality in any species has to be clearly understood in order to conduct rational genetic studies as well as breeding programmes. Mushrooms representing three types with respect to life cycle and genetic controls for mating have been considered. It is hoped that the examples discussed in this course will provide some basic and useful guidelines for procedures as they might be applied generally to any breeding programme for each type of mushroom. Although it seems likely that selection from existing commercial strains of edible mushrooms may continue to enjoy some improvement, by analogy with other crop species, hybridisation seems to offer the best prospects for real progress, especially with regard to multiple gene transfers mediated by protoplast fusion, during the next few decades. Finally, the experience of breeding in a variety of organisms has demonstrated the wisdom of preserving the starting strains used in all breeding programmes for the purpose of maintaining a pool of genetically diverse individuals. Such a practice would allow the possibility of carrying out additional breeding programmes to either improve or restore strains, which are in current use.

MUSHROOM CULTIVATION

The cultivation of mushrooms can be both a relatively primitive farming activity, and a high technology industry. In each case, however, continuous production of successful crops requires both practical experience and scientific knowledge. Mushroom cultivation is both a science and an art. The science is developed through research; the art is perfected through curiosity and practical experience.

Mushroom growth dynamics involve some technological elements, which are in consonance with those exhibited by our common agricultural crop plants. For example, there is a vegetative growth phase, when the mycelia grow profusely; and a reproductive (fruiting) growth phase, when the umbrella-like body that we call mushroom develops. In the agricultural plants, e.g., sunflowers, when the plants switch from the vegetative growth to the reproductive growths, retarded tips for further growth (elongation) is an obvious phenomenon of mature. It is the same principle in mushroom production. After the vegetative (mycelial) phase has reached maturity, what the mushroom farmer needs next is the induction of fruiting. This is the time the mycelial growth tips should be retarded by regulating the environme... factors. These factors generally called "triggers" or "environmental shocks", such as, switching on the light, providing fresh air, and lowering temperatures, can trigger fruiting.

Although the principles of cultivation are commonly similar for all mushrooms, the practical technologies can be quite different for different species cultivated. The technologies have to be modified and adjusted according to the local climatic conditions, materials available for substrates and varieties of the mushroom used.

Figure 3. Lentinula edodes grown on sawdust synthetic logs

The outlook for many of the known mushroom species is bright. Production of mushrooms worldwide has been steadily increasing, mainly due to contributions from developing countries, such as China, India, Poland, Hungary and Vietnam. There are also increasing experimentally based evidence to support centuries of observations regarding the nutritional and medicinal benefits of mushrooms. The value of mushrooms has recently been promoted to tremendous levels with medicinal mushrooms trials conducted for HIV/AIDS patients in Africa, generating encouraging results. However, harvests of highly prized edible mycorrhizal mushrooms are continuously decreasing. This has triggered research into devising methods for improved cultivation of wild mushroom. It is hoped that there will be even more research into this area, so that larger quantities of wild mushrooms can be massively harvested through artificial cultivation or semi-cultivation methods. Technological developments in the mushroom industry in general has seen increasing production capacities, innovations in cultivation technologies, improvements to final mushroom goods, capitalising on mushrooms' nutritional and medicinal properties, and utilising mushrooms' natural qualities for environmental benefits.

Mushroom farming is a complex business, which requires precision. Indeed, it is not as simple as what some people often loosely stipulate. It calls for adherence to precise procedures. The major practical steps/segments of mushroom cultivation are: (a) selection of an acceptable mushroom species; (b) secreting a good quality fruiting culture; (c) development of robust spawn; (d) preparation of selective substrate/compost; (e) care of mycelial (spawn) running; (f) management of fruiting/ mushroom development; and (g) harvesting mushrooms carefully. If you ignore one critical step/segment, you are inviting trouble, which could lead to a substantially reduced mushroom crop yield and mushroom marketing value.

Selection of Acceptable Mushroom Species/Strains

Before any decision to cultivate a particular mushroom is made, it is important to determine if that species possess organoleptic qualities acceptable to the indigenous population or to theinternational market, if suitable substrates for cultivation are plentiful, and if environmental requirements for growth and fruiting can be met without excessively costly systems of mechanical control.

Secreting a Good Quality Fruiting Culture

A "fruiting culture" is defined as a culture with the genetic capacity to form fruiting bodies under suitable growth conditions. The stock culture which is selected should be acceptable in terms of yield, flavour, texture, fruiting time, etc.

Sources of the cultures

Tissue culture. A large healthy mushroom should be chosen either in the later button or egg stage. It should be cleaned with 75% alcohol. The mushroom should be split in half by hand longitudinally and some inside tissue taken from the upper part of the stipe. It should be placed centrally on the surface of the medium with a sterilised needle. The quicker this is done the better. As soon as we transfer the tissue, the test tube should be closed and dated before it is returned to the incubator between 25°C and 34°C depending on the mushroom used. Within two or three days some white, delicate mycelia will be produced from the small piece of the tissue. They grow upwards encircling the inner wall of the test tube. About ten days later the mycelium will grow rapidly and cover the surface of the agar medium. Then it is ready to transfer to spawn substrate to make spawn.

Spore culture. Individual spores properly collected can be transferred singly to a testtube or petri dish and allowed to develop and germinate into mycelium. Some single-spore isolates from homothallic mushrooms, e.g. Volvariella volvacea (primary homothallism) or Agaricus bisporus (secondary homothallism) can be used as fruiting culture to make spawn. However if single-spore isolates are from heterothallic mushrooms, e.g. Lentinula edodes, Pleurotus sajor-caju and Ganoderma lucidum, then they cannot form fruiting cultures and thus cannot make spawn. They have to be mated with a compatible single-spore isolate. After mating they form a dikaryon/fruiting culture. Then they can be used to make spawn.

Pure culture from other laboratories. As an alternative to culturing in the laboratory, as outlined, a test tube culture may be obtained from a research laboratory. The advantage of this is that cultures maintained in reputable culture collections are already tested for their production characteristics and are guaranteed to be pure.

Cultures from another source. Cultures also may be grown from spawn obtained from another source. A piece of the spawn is aseptically transferred to agar slants. However, this is risky because the number of transfers that the spawn culture has undergone is rarely known. If this procedure is followed, it may be advisable to first grow the spawn into fruiting bodies, then make the necessary isolations from the fruiting body.

Culture media

Mushrooms grow on a variety of culture media and on different agar formulas, both natural and synthetic, depending on the organism to be cultivated and the purpose of the cultivation. Synthetic media are often expensive and time-consuming in preparation hence they are not commonly used for routine purposes in mushroom laboratories.

PDA (potato dextrose agar), is the simplest and the most popular medium for growing mycelia of most cultivated mushrooms. It can be purchased commercially as ready mixed powder which can be used directly to make the medium in the laboratory, with a concentration of 20gm/1 litre of distilled water.

Alternatively, it can be prepared in the laboratory with the following ingredients: Potato, diced - 200 gm (1/5kg); Dextrose - 20 gm; Powdered agar or agar bars - 20gm; Water - 1 litre. Procedure: Peeled potatoes are washed, weighed, and cut into cubes. They are boiled in a casserole with at least one litre of water until they become soft (around 15 minutes). The potatoes are removed and water is added to the broth to make exactly 1 litre. The broth is returned to the casserole, and dextrose and the agar added. The solution is heated and stirred occasionally until the agar is melted. The hot solution is then poured into clear flat bottles filling to about 2.5cm from the bottom. When using test tubes for the stock cultures, they are filled with at least 10ml of the liquid agar solution. The bottles or test tubes are plugged with cottonwool. When Petri dishes are available and are used to produce mycelial colonies, the solution can be poured into the dishes to form a layer on the bottom.

A ready-made MEA (malt extract agar) powder is also available commercially. The recommended amount of powder (20gm) is mixed with 1 litre of water, then melted and sterilised. One percent peptone or 0.5% yeast may be added for faster mycelial growth for both PDA and MEA.

Development of Robust Spawn

A medium through which the mycelium of a fruiting culture has grown and which serves as the inoculum of "seed" for the substrate in mushroom cultivation is called the "mushroom spawn". Failure to achieve a satisfactory harvest may often be traced to unsatisfactory spawn used. Consideration must also be given to the nature of the spawn substrate since this influences rapidity of growth in the spawn medium as well as the rate of mycelial growth and filling of the beds following inoculation.

Mushroom Spawn Preparation

Definitions of spawn and spawning. The word "spawn" is derived from an old French verb, espandre, meaning to spread out or expand, which was derived from the Latin, expandere, meaning to spread. Spawn is also defined by Webster's Dictionary as "the mycelium of fungi, especially of mushrooms grown to be eaten, used for propagation". In the mushroom industry, spawn is a substrate into which a mushroom mycelium has impregnated and developed, and which will be used as a seed in propagation for mushroom production. In addition the verb, to spawn, is used to mean inoculation of a substrate with mushroom spawn.The simple definition of spawning is the planting of mushroom spawn in the prepared compost/substrate. Along with advances in spawn making, the methods of spawning have also been continuously developed and improved, making it possible for the mushroom mycelium to grow through the compost more quickly.

Figure 4. Fruiting bodies of the Straw mushroom grown on cotton waste as substrate

Spawn Substrates. A number of materials, mostly agricultural wastes, can be used to prepare mushroom spawn. The type of waste available varies from region to region. Some of these wastes are chopped rice straw, sawdust, water hyacinth leaves, used tea leaves, cotton wastes and lotus seed husks. In most laboratories, cereal grains (wheat, rye or sorghum) are used as mother spawn, and agricultural wastes as the planting spawn substrates. The mother spawn is used to inoculate the final spawn container in which the planting spawn will be produced. The planting spawn is used to inoculate the mushroom cultivation compost/substrate for fruiting/mushroom production.

Preparation of mother spawn. Here using wheat grains as an example, they are soaked in water for 2 hours or over night. Dead seeds or those that float on water should be carefully removed. Then the grains are washed again and boiled in water for at least 10 to 15 minutes until they expand but not quite broken. The grains are drained and allowed to cool. Precipitated chalk (1.5% on wet basis) is added to the grains. The grains are then loosely packed in bottles which are 2/3 full. These are plugged with cotton wool or covered by double-layered aluminium foil. The grains are sterilised in a pressure cooker for about 1 hour at 121°C, alternatively, they are steamed for 3 to 4 hours in a large cast-iron casserole. The bottles are then cooled prior to inoculation.

Preparation of planting spawn. Here we shall use rice straw or water hyacinth leaves as an example for Pleurotus sajor-caju planting spawn. The rice straw (or water hyacinth leaves) is chopped into pieces about 2 to 3 cm (1 inch) long, then soaked in water for 4 - 12 hours. The excess water is drained off and the straw pieces mixed with a solution of 1% sucrose, 1.5% chalk and 2% wheat/rice bran in water. The final moisture content is to be adjusted to about 60%. The mixture is then put into glass bottles or plastic bags and sterilised for at least one hour at 121°C before being inoculated with the mother spawn.

Mushroom Spawn Handling

Maintenance of spawn quality. Mushroom spawn, whether prepared as a family home project or on industrial scale using modern equipment, should be in excellent condition when delivered to growers. Spawn of most mushrooms can be refrigerated, but it should be warmed to normal room temperature before it is used as an inoculum or as planting spawn. Vigorous growth of the planting spawn is a prerequisite to good growth and yield. If the spawn is not vigorous, the mushroom mycelium will be overgrown by competitor organisms. If it is vigorous it will overcome many of the competitive organisms and produce more mushrooms. When purchasing spawn, ask the spawn maker how long the spawn can be kept before planting. Old spawn is not acceptable because its vigour may have decreased. Buyers or users should know the "expiration" date of the spawn.

Spawn Quantities. The quantity of the spawn used does not directly affect yield. However, the use of more spawn may reduce the effect of competitive organisms present in the planting substrates. The greater the amount of spawn used, the faster it will colonise the substrate. As a result, the growth of competitors is hindered, and yield will be regular and not affected by this competition. 2-4% of spawn is suggested to be inoculated into the spawning substrate. Once the container is opened, spawn should be used in its entirety. Unused and opened bottles or bags of spawn, however, can be kept in the refrigerator for 2 to 6 days as long as they are not contaminated (i.e., no unwanted fungi are growing on the surface) during storage.

Notes on commercial spawn supplies. When growing mushrooms on a small scale, it is not necessary to prepare your own spawn. Commercial suppliers of spawn who provide material to small growers are usually available. Spawn should be ordered ahead of time so that it will be of the right age. Contaminated, old or no-growth spawn should never be sold to growers. Spawn makers should maintain a testing facility where they can test each batch of spawn for production characteristics. Sales personnel should visit growers using their spawn so that theycan observe problems at first hand. Problems related to spawn production must be corrected quickly.

Preparation of Selective Substrate/Compost

While a sterile substrate free from all competitive micro-organisms is the ideal medium for cultivating edible mushrooms, systems involving such strict hygiene are generally too costly and impractical to operate on a large scale. Substrates for cultivating edible mushrooms normally require varying degrees of pre-treatment in order to promote growth of the mushroom mycelium to the practical exclusion of other micro-organisms. The substrate must be rich in essential nutrients in forms which are readily available to the mushroom, and be free of toxic substances which inhibit growth of the spawn. Moisture content, pH and good gaseous exchange between the substrate and the surrounding environment are important physical factors to consider. (1). Mushroom substrate. Mushroom substrate may be simply defined as a kind of lignocellulosic material which supports the growth, development, and fruiting of mushroom mycelium. The process of preparation of substrate is broadly termed "composting". The final product of "composting" is called the "compost" or prepared substrate. The process for preparation of substrates has been the subject of much scientific and practical interest over the past two decades. It should be noted that different types of mushrooms require different types or substrate/compost.

Agaricus bisporus grows on fermented compost which is traditionally developed from wheat straw mixed with horse manure, and it requires higher nitrogen content. Wheat straw contains about 0.62% nitrogen, and horse manure contains about 1.5 to 1.8% nitrogen. The optimum C:N (carbon:nitrogen) ratio for the mushroom is about 17 to 1 for mycelial running. Volvariella volvacea, the paddy straw mushroom, and Stropharia rugosoannulata are grown on almost raw or less composted plant residues, such as rice straw and cotton waste materials. These cellulosic plant materials contain different amounts of nitrogen: rice straw 0.58%, cotton waste 0.65 to 1%, and banana leaves 1.71%. It should be understood that V. volvacea is capable of growing on plant material with low nitrogen content. The optimum C:N ratio is about 75 to 80:1, but C:N ratios from 32 to 150:1 are almost as effective. Lentinula edodes and Pleurotus spp. are fungi that can grow on wood. In addition to being distinguished by its high lignin content, wood can also be distinguished from other plant materials by its very low nitrogen content.

Woody tissues contain 0.03 to 1.0% nitrogen as compared to 0.85 to 1.71% in herbaceous residues. The C:N ratio in most woody tissues is in the order of 350 to 500:1. Woodinhabiting mushrooms are unique in that they can grow in such substrates. This suggests that these mushrooms can metabolise large amounts of carbohydrates, including lignin, in the presence of a very small amount of nitrogen.

Composting. Some early composting investigators had a good appreciation of the biological nature of the composting process. These early researchers had some appreciation of the role of various ecological factors such as temperature, oxygen, moisture, pH and nutrition in determining the general activity and population dynamics that occur during composting. As a working hypothesis, it may be assumed that composting conditions which produce a favourable medium for the development of mushrooms probably do so because they encourage the development of microbial population that is best able to pave the way for the subsequent growth and fructification of mushrooms. Such a hypothesis must take full cognisance of the effect of thestaling products of different groups of organisms on the mushroom development as well as the actions of these organisms in producing changes in the manure favourable to the nutrition of mushrooms under competitive conditions.

The substrate left after the mushrooms have been harvested is known as spent compost. This is present in large amounts, and raises the question of what can be done with it. It is certainly not desirable to leave it as a possible source of pollution. It is known that there still remains in the spent compost a considerable amount of lignocellulosic material in addition to the mushroom mycelia and also other products formed by the metabolic activities of the mycelium. Thus, the spent compost should be capable of supporting further biological activities, e.g., the growth of another species of edible mushroom; use as fodder for livestock; as a soil conditioner and fertiliser; and also in bioremediation.

Care of Mycelial (Spawn) Running

Following composting, the substrate is placed in beds where it is generally pasteurized by steam to kill off potential competitive microorganisms. After the compost has cooled, the spawn may be broadcast over the bed surface and then pressed down firmly against the substrate to ensure good contact, or inserted 2 to 2.5 cm deep into the substrate. Spawn running is the phase during which mycelium grows from the spawn and permeates into the substrate. Good mycelial growth is essential for mushroom production.

Management of Fruiting/Mushroom Development

Under suitable environmental conditions, which may differ from those adopted for spawn running, primordial formation occurs and then followed by the production of fruiting bodies. The appearance of mushrooms normally occurs in rhythmic cycles called "flushes".

Harvesting Mushrooms Carefully

Harvesting is carried out at different maturation stages depending upon the species and upon consumer preferences and market value.

MUSHROOM DISEASES

The mushroom, like any other cultivated crop, is subject to attack by pathogens and pests. The mushroom diseases can be caused by both fungi and bacteria. There are four important fungal diseases of the cultivated mushrooms, particularly referring to Agaricus bisporus and these are:

Dry Bubble caused by Verticillium Fungicola

At vary early stages on infection, brown spots can be seen on mushroom caps. This stage of the disease can be confused with the brown blotch disease caused by a bacterium. However, in the case of dry bubble, if the infected mushrooms continue to be incubated in a humid chamber, greyish white mycelium of the pathogen will develop from the brown spots on the mushroom caps.

Control: The environment conditions most conducive to the growth of Verticillium - high relative humidity and high relative temperature. Therefore, an effective ecological control would be to reduce the relative humidityand temperature in the growing rooms. A reduction in relative humidity from 90-95% to 80-85% and in temperature to 14o C can result in a reduction in the incidence of the disease. Control of the disease depends also on farm hygiene consisting of prompt disposal of spent compost and debris and disinfection of affected bed areas. If soil is used as a casing material, it should be pasteurized at 60oC for half an hour with aerated steam.

Wet Bubble caused by Mycogone permiciosa

Infection of mushroom fruit bodies results in malformed stalk and cap. The surface of diseased mushrooms is covered by a white felt of mycelium of the pathogen. The undifferentiated tissues of the cap and stalk become necrotic with a wet soft rot dripping a brown liquid. The name "wet bubble" is derived from the foul-smelling, mis-shapen wet mushroom tissues.

Control: Disinfection of casing material or steam-air pasteurization of soil, if soil is used as a casing material, can control the disease. It is necessary to protect the developing mushroom fruiting bodies throughout the period of cropping. It is essential that management gives full attention to hygiene such as disinfecting the work area, removing all spent materials, disinfecting all harvesting tools and carrying out after-crop sterilization using steam at 70oC for 12 hours.

Mildew caused by Cladobotrym sp.

This disease is characterized by a cotton wool type of growth of the pathogen on the mushroom fruiting bodies and spreads over the casing layer to neighbouring mushrooms. The greyish-white mycelium of the pathogen can envelop the mushroom fruiting body completely. Infected mushrooms die with a soft wet rot.

Control: Disinfection of the casing material and strict hygiene can control the disease.

Aphanocladium

Disease caused by Aphanocladium album. Infected mushrooms develop brown spots which, under conditions of high relative humidity, show greyish-white mycelium of the pathogen. The gill tissues of the mushroom fruiting body are commonly infected by the pathogen.

Control: Disinfection of the casing material appears to control the disease. Management of the crop environment similar to that of the Dry and Wet Bubble diseases can be effective in controlling the disease. In addition, there are three important bacterial diseases of the cultivated mushrooms, particularly referring to Agaricus bisporus:

Brown Blotch caused by Pseudomonas Tolaasii

The disease causes brown, slightly sunken blotches on the developing mushroom fruiting body. It affects only the surface layers of the mushroom and, in severe cases, brown streaks develop on the stalk. The blotches on the cap can, sometimes, be yellowish to pale brown in colour.

Control: The disease can be controlled by manipulating the crop environment. Since the pathogen can spread by splashing water, the surfaces of mushrooms should be maintained relatively dry. If the surfaces of mushrooms dry within an hour or two after becoming wet, no infection occurs. The drying can take place in an atmosphere of high relative humidity as long as there is no fluctuation in temperature and there is a circulation of air over the mushroom beds.

Mummy

Disease caused by Pseudomonas sp. In diseased mushrooms the stalk is bent and the cap tilted. Often there is a dense growth of mycelium around the base of the stalk on the surfaceof the casing layer. Infected mushrooms remain intact for a relatively long period after which they are invaded by secondary bacteria. Mushrooms often fail to mature and remain in the "button" stage with unopened veil.

Control: Strict hygiene is the best means of controlling the disease. Disinfecting the casing layer can help to reduce the infection.

Drippy Gill caused by Pseudomonas agarici

The pathogen damages the gill tissues of mature mushroom fruiting bodies. Damage is restricted to the gills. Dark brown, round spots appear on the side and bottom edges of the gills. At the centre of each spot, a creamy grey droplet may show on the gill surface. If the infection is severe, bacterial droplets join to form long streaks of slime which may lead to distortion and collapse of the gills.

Control: Disinfection of casing soil can control the disease.

Post-Harvest Handling

Like all fruits and vegetables, mushrooms are perishable, and after harvest they often change in ways that make them unacceptable for human consumption. The most readily observable of these changes include wilting, ripening, browning, liquefaction, loss of moisture, and loss of texture, aroma and flavour. To ensure that mushrooms are acceptable and nutritious to the consumer at the time of purchase, it is necessary to delay or prevent senescence. Expansion of the pileus by growth of gills and elongation of the stipe post harvest is supported by increased cell wall chitin and protein. Chitin synthase can be activated by proteinases. It has shown that during ageing, there is a major redistribution of dry weight between tissues with gill dry weight increasing during storage while that of the pileus and stipe diminishing. The rate of cap opening depends on the stipe length (the longer the stipe, the greater the expansion), indicating that the stipe is acting as a major nutrient source for the expanding gill tissue.

Recently, the biochemistry and genetic regulation in mushrooms post harvest has been studied by identifying genes h higher transcript levels after harvest. Technologies such as cooling and modified atmosphere packaging can be use to delay the rate of senescence, while preservative technologies such as canning, drying, pickling, and freezing and _-irradiation arrest biological function to prevent senescence. Depending on the species, the shelf life of mushrooms may vary from one day to two weeks. Fresh mushrooms are best-stored unwashed in brown paper bags in the refrigerator, preferably on the lowest shelf. Bags should be available from your mushroom retailer; otherwise, a paper lunch bag is fine. Generally, it is important that fresh mushrooms are packaged in materials that allow them to breathe, so they do not 'sweat' and become slimy. At the same time, the material should ensure mushrooms do not dry out too much.

REFERENCES

Elliott, T. J. 1982. "Genetics and breeding of cultivated mushrooms". In *Tropical Mushrooms*, edited by S. T. Chang and T. H. Quimio. The Chinese University Press, Hong Kong.

Hall I. R., Y. Wang and A. Amicucci. 2003. "Cultivation of edible ectomycorrhizal mushrooms". *Trends in Biotechnology* 21, 433-438.

Kaul, T. N. and B. L. Dhar. 2007. *Biology and Cultivation of Edible Mushrooms*. Westville Publishing House. New Delhi, 240pp

Miles, P. G. and S. T. Chang. 1997. *Mushroom Biology: Concise Basics and Current Developments*. World Scientific. Singapore, 194p.

17

Turnip

The turnip (*Brassica rapa* var. rapa) is a root vegetable commonly grown in temperate climates worldwide for its white, bulbous taproot. Small, tender varieties are grown for human consumption, while larger varieties are grown as feed for livestock.The most common type of turnip is mostly white-skinned apart from the upper 1-6 centimeters, which protrude above the ground and are purple, red, or greenish wherever sunlight has fallen. This above-ground part develops from stem tissue, but is fused with the root. The interior flesh is entirely white. The entire root is roughly conical, but can be occasionally tomato-shaped, about 5-20 centimeters in diameter, and lacks side roots. The taproot is thin and 10 centimeters or more in length; it is trimmed off before marketing. The leaves grow directly from the above-ground shoulder of the root, with little or no visible crown or neck.

Turnip leaves are sometimes eaten as "turnip greens", and they resemble mustard greens in flavor. Turnip greens are a common side dish in southeastern US cooking, primarily during late fall and winter. Smaller leaves are preferred; however, any bitter taste of larger leaves can be reduced by pouring off the water from initial boiling and replacing it with fresh water. Varieties specifically grown for the leaves resemble mustard greens more than those grown for the roots, with small or no storage roots. Varieties of B. rapa that have been developed only for use as leaves are called Chinese cabbage. Both leaves and root have a pungent flavor similar to raw cabbage or radishes that becomes mild after cooking.

Turnip roots weigh up to about 1 kilogram, although they can be harvested when smaller. Size is partly a function of variety and partly a function of the length of time that the turnip has grown. Most very small turnips are specialty varieties. These are only available when freshly harvested and do not keep well. Most baby turnips can be eaten whole, including their leaves. Baby turnips come in yellow-, orange-, and red-fleshed varieties as well as white-fleshed. Their flavor is mild, so they can be eaten raw in salads like radishes.

Turnips are high only in Vitamin C. On the other hand, turnip greens are a good source of Vitamin A, folate, Vitamin C, Vitamin K and calcium. Turnip greens are high in lutein (8.4mg / 100g).

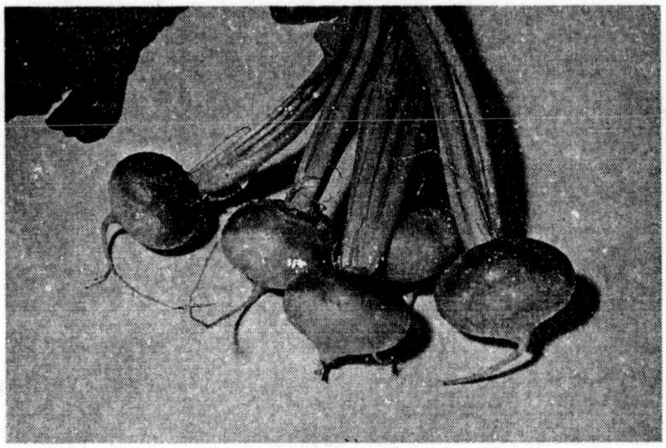

Figure 1. Turnips

The turnip was a well-established crop in Hellenistic and Roman times, which leads to the assumption that it was brought into cultivation earlier. But Zohary and Hopf note that "there are almost no archeological records available" to help determine its earlier history and domestication. Wild forms of the hot turnip and its relatives the mustards and radish are found over west Asia and Europe, suggesting that their domestication took place somewhere in that area. However Zohary and Hopf conclude, "Suggestions as to the origins of these plants are necessarily based on linguistic considerations."

Turnip (*Brassica rapa* L.) is a root Brassica crop and has been used as a vegetable for human consumption in Europe since prehistoric times. Turnip root has been a popular livestock fodder for at least 600 years wherever the crop can be grown. For most of that time turnip roots have been managed as forage. Researchers in the United States determined in the early 1900s that turnip roots are valuable energy sources for young ruminant animals. However, livestock farmers at that time were turning away from the Brassica root crops (which also include rutabagas or swedes) for fodder because much hand labor was required for the production and utilisation of the large roots. One study showed that the labor requirement on a nutrient basis for these crops was three times that needed for corn silage production.

In the late 1970s, however, researchers began to demonstrate the potential of turnip as pasture. The development of varieties with partially exposed roots rendered the roots more available to grazing animals. Livestock graze turnip tops and roots readily, and the forage is of high quality. Pasturing eliminates the need for manual labor in harvesting and storing. In general, the root Brassicas are fast-growing, high yielding and well adapted to seeding into existing pastures with little or no tillage or to seeding into a conventionally prepared seedbed.

Turnip is a cool-weather crop and well adapted for the northern parts of the United States and Europe and for Great Britain and Canada. However, truck-growing areas of the South also produce turnip roots and greens in all seasons for human consumption.

Turnip produces high-quality forage if harvested before heading. Livestock eat the stems, leaves and roots of turnip plants. Above-ground parts normally contain 20 to 25% crude protein, 65 to 80% in vitro digestible dry matter (IVDDM), about 20% neutral detergent fiber (NDF) and about 23% acid detergent fiber (ADF). The roots contain 10 to 14% crude protein and 80 to 85% IVDDM.

The high levels of glucosinolates (which can cause thyroid enlargement in young growing sheep and cattle) can be a problem if turnip forage is fed for long enough. Glucosinolates are higher in older (90-day) than 60-day forage. Oral or subcutaneous iodine administration can alleviate thyroid problems. Turnip roots usually are higher in glucosinolates than the tops and leaves. Two other anti-quality factors, S-methyl cysteine sulphoxide (SMCO) and free nitrates can also be present. SMCO is the main problem and can cause anemia. To minimise the potential for animal health problems from these factors, forage from turnips should be fed in combination with other forages.

Turnip and other Brassicas can provide grazing at any time during the summer and fall depending on the seeding date. A promising use may be for late fall grazing. These crops maintain their forage quality, if not headed, well into the fall even after freezing temperatures and may be grazed in the Upper Midwest into November. Many turnips can be grazed twice to permit utilisation of top growth and roots.

CULTIVATION

The benefits derived from turnip husbandry are of great magnitude; light soils are cultivated with profit and facility; abundance of food is provided for man and beast; the earth is turned to the uses for which it is physically calculated, and by being suitably cleaned with this preparatory crop, a bed is provided for grass seeds, wherein they flourish and prosper with greater vigor than after any other preparation.

The first ploughing is given immediately after harvest, or as soon as the wheat seed is finished, either in length or across the field, as circumstances may seem to require. In this state the ground remains till the oat seed is finished, when a second ploughing is given to it, usually in a contrary direction to the first. It is then repeatedly harrowed, often rolled between the harrowings and every particle of root-weeds carefully picked off with the hand; a third ploughing is then bestowed, and the other operations are repeated. In this stage, if the ground has not been very foul, the seed process.

The next part of the process is the sowing of the seed; this may be performed by drilling machines of different sizes and constructions, through all acting on the same principle. A machine drawn by a horse in a pair of shafts, sows two drills at a time and answers extremely well, where the ground is flat, and the drills properly made up. The weight of the machine ensures a regularity of sowing hardly to be gained by those of a different size and construction. From two to three pounds of seed are sown upon the acre (2 to 3 kg/hectare), though the smallest of these quantities will give many more plants in ordinary seasons than are necessary; but as the seed is not an expensive article the greater part of farmers incline to sow thick, which both provides against the danger of part of the seed perishing, and gives the young plants an advantage at the outset.

Turnips are sown from the beginning to the end of June, but the second and third weeks of the month are, by judicious farmers, accounted the most proper time. Some people have sown as early as May, and with advantage, but these early fields are apt to run to seed before winter, especially if the autumn be favorable to vegetation. As a general rule it may be laid down that the earliest sowings should be on the latest soils; plants on such soils are often long before they make any great progress, and, in the end, may be far behind those in other situations, which were much later sown. The hot turnip plant, indeed, does not thrive rapidly till its roots reach the dung, and the previous nourishment afforded them is often so scanty as to stunt them altogether before they get so far.

The first thing to be done in this process is to run a horse-hoe, called a scraper, along the intervals, keeping at such a distance from the young plants that they shall not be injured; this operation destroys all the annual weeds which have sprung up, and leaves the plants standing in regular stripes or rows. The hand hoeing then commences, by which the turnips are all singled out at a distance of from 8-12 inches, and the redundant ones drawn into the spaces between the rows. The singling out of the young plants is an operation of great importance, for an error committed in this process can hardly be afterward rectified. Boys and girls are always employed as hoers; but a steady and trusty man-servant is usually set over them to see that the work is properly executed.

In eight or ten days, or such a length of time as circumstances may require, a horse-hoe of a different construction from the scraper is used. This, in fact, is generally a small plough, of the same kind with that commonly wrought, but of smaller dimensions. By this implement, the earth is pared away from the sides of the drills, and a sort of new ridge formed in the middle of the former interval. The hand-hoers are again set to work, and every weed and superfluous turnip is cut up; afterward the horse-hoe is employed to separate the earth, which it formerly threw into the furrows, and lay it back to the sides of the drills. On dry lands this is done by the scraper, but where the least tendency to moisture prevails, the small plough is used, in order that the furrows may be perfectly cleaned out. This latter mode, indeed, is very generally practiced.

Growth Habits

Turnip is a member of the mustard family and is therefore related to cabbage and cauliflower. Turnip is a biennial which generally forms seed the second year or even late in the fall in the first year if planted early in the spring. During the first or seeding year 8 to 12 erect leaves, 12 to 14 in. tall with leaf blades 3 to 5 in. wide are produced per plant. Turnip leaves are usually light green, thin and sparsely pubescent (hairy). In addition, a white-fleshed, large global or tapered root develops at the base of the leaf petioles. The storage root varies in size but usually is 3 to 4 in. wide and 6 to 8 in. long. The storage root consists mainly of the hypocotyl, the plant part that lies between the true root and the first seedling leaves (cotyledons). The storage mot generally has little or no neck and a distinct taproot. The storage root can overwinter in areas of mild winter or with adequate snow cover for insulation and produce 8 to 10 leaves from the crown in a broad, low-spreading growth habit the following spring. Branched flowering stems 12 to 36 in. tall are also produced. The flowers are clustered at the top of the raceme and are usually raised above the terminal buds. Turnip flowers are small and have four light-yellow petals.

Environment Requirements

Climate

Brassicas are both cold-hardy and drought-tolerant. They can be planted late-even as a second crop-and provide high-quality grazing late in the fall. Turnip planted in July will provide grazing from September to November. The most vigorous root growth takes place during periods of low temperature (40 to 60¡F) in the fall. The leaves maintain their nutritional quality even after repeated exposure to frost.

Soil

Like other Brassicas, turnip grows best in a moderately deep loam, fertile and slightly acid soil. Turnip does not do well in soils that are of high clay texture, wet or poorly drained. For good root growth turnip needs a loose, well aerated soil.

Seedbed Preparation

Turnip seed is small and it is essential that it be seeded into a fine, firm seedbed with adequate moisture for germination. Plow and disk or harrow to produce a seedbed that is fine, firm and free of weeds and clods.

Turnip, like other Brassicas, can also be seeded into a sod or into stubble of another crop with minimum tillage. When seeding into sod, it should be suppressed or killed, as the young Brassica seedlings cannot compete with established grasses. To kill sod, apply 2 qt/acre of Roundup at least three days prior to seeding. A 0.5 qt/acre rate of Roundup can be used in 3 to 10 gal water/acre to suppress sod or to prepare a field of wheat stubble for seeding with turnip. Once established, turnip will compete with most weeds.

The advantages of direct drilling turnip into sod include fewer crop losses due to insect pests, such as the flea beetle, and less soil erosion on sloping sites where pastures are often located. A field of turnip established in sod gives animals a firm footing in all kinds of weather. It also allows the original sod to grow again the following spring if it has only been suppressed.

Seeding Dates

Turnip seed does not germinate well in cold soil. Turnip should not be planted until the soil temperature is at least 50¡F or at corn planting time. The crop can be planted any time during the summer until about 70 days before a killing frost. Plantings after these dates may not have sufficient time to produce good forage growth.

Method and Rate of Seeding

Turnip seed can be planted in 6 to 8 in. rows at a rate of 1.5 to 2.5 lb/acre with a minimum-till drill when sod seeding. In conventionally prepared seedbeds, the crop can be seeded with a forage crop seeder or broadcast followed by cultipacking. The seed should not be covered with more than 1/2 in. of soil. A plant population of 5 to 6 per sq. ft. is desirable.

Fertility and Lime Requirements

Good soil fertility is very important for good yields. Soil tests should be taken to assure proper fertilisation. Lime acid soils to pH 6.0. Fertilisers should be applied at the time of seeding or within 3 days of seeding to give the crop a competitive edge on weeds. Apply 100 lb/acre nitrogen to soils containing 2 to 5% organic matter, 120 lb/acre if less than 2% organic matter and 60 to 80 lb/acre if more than 5% organic matter. Requirements for phosphorus and potassium are similar to those of a small grain. In Wisconsin and Minnesota, when soil tests are in the medium range, about 20 to 30 lb/acre of P_2O_5 and 120 lb/acre of K_2O should be applied. Fertiliser applications should be banded at least 2 in. to the side and below the seed or broadcast. Boron and sulfur may also be needed. If the soil tests "low" in boron, apply 1 lb boron/acre on sandy soils, and twice this amount on other soils. Apply 10 to 15 lb of S/acre if a soil sulfur test indicates a need for this element.

Variety Selection

Three forage turnip varieties are recommended for use in the Upper Midwest: Green Globe, and York Globe from New Zealand and Sirius turnip from Sweden. In Pennsylvania, Green Globe and York Globe yielded more than Sirius at 60 days after planting, but Green Globe reached its peak yield later than the other two. Sirius yields were more variable from year to year than Green Globe or York Globe. The tops and leaves of Sirius have less glucosinolate than the other two varieties.

Weed Control

Weeds are generally not a problem once the turnip crop is established. However, sod and annual weeds should be controlled chemically and/or culturally before planting. Sod can be suppressed or killed with Roundup, as described under Seedbed Preparation. If annual weeds are present at planting time, eliminate them with a burndown herbicide such as Gramoxone. Tillage before planting can be used for weed control on a conventional seedbed.

Diseases and their Control

Turnip crops may suffer from clubroot, root knot, leaf spot, white rust, scab, anthracnose, turnip mosaic virus and rhizoctonia rot. in some cases, diseases can lead to crop failure if rotation or other control measures are not used. Resistant varieties are available for some diseases. To prevent problems with diseases, Brassicas should not be grown on the same site more than two years in a row. If clubroot is a problem, rotation should be six years.

Insects and Other Predators and their Control

Turnip crops are attacked by two different flea beetles, which eat holes in the cotyledons and first leaves, chew stems and cause extensive plant loss. The cabbage flea beetle and the striped flea beetle feed exclusively on Brassicas, including related weeds such as yellow rocket. Problems with these flea beetles are much greater when Brassicas are grown under conventional tillage. Both flea beetles can be controlled with insecticides applied to the soil at planting.

Turnip crops can also be damaged by infestations of the common turnip louse or aphid. This insect feeds on the undersides of the leaves and may be so close to the ground that it is difficult to reach with a dust or a spray. In cases of severe infestation, the outer leaves curl and turn yellow. Aphid-tolerant varieties such as 'Forage Star' can give some protection against this insect.

Harvesting

Turnip plants are ready for grazing or green-chop when the forage is about 12 in. tall (70 to 90 days after planting). It is best not to wait too long because fungal diseases may begin to cut yields approximately 110 days after planting. The pasture should be grazed for a short time and the livestock removed to allow the plants to regrow. If grazed to a 5 in. stubble, 1 to 4 grazing periods may occur, depending on planting date and growing conditions. Strip or block-grazing is desirable to insure complete grazing.

The forage quality of turnip is sufficiently high, especially in protein, that it should be considered similar to concentrate feeds, and precautions should be taken to prevent animal health problems. Livestock should not be hungry when put on pasture the first time so they do not gorge themselves. If the livestock are moving from a feed low in nutritional value, feed a high-quality diet for two to three weeks prior to grazing turnip, or feed turnip for 30 min/day for one week prior to heavier grazing. This will allow for the development of a rumen microbial population that is adequate to digest the high levels of protein in forage turnips. A lower quality hay should be made available (2 to 3 lb of dry roughage/head/day for sheep and 10 to 15 lb for cattle) to provide some fiber in the animals' diet.

Yield Potential and Performance Results

Yields of forage turnip range between 3 and 4 tons of dry matter/acre when harvested or grazed about 90 days after planting. Up to 1,000 grazing days/acre for 900 lb steers and 2,300 grazing days/acre for 90 lb lambs have been obtained for Forage Star turnip. Turnip is a highly nutritious forage crop that has a short growing season and can provide late fall grazing after other forage crops are finished for the year. Seed prices range from $1.50/lb for garden-variety turnips to about $8/lb for some new hybrid varieties.

REFERENCES

Daniel Zohary and Maria Hopf, 2000. *Domestication of plants in the Old World*, third edition Oxford: University Press.

Hartemink, A.E. 2001. *Soil fertility in the tropics with case studies on plantation crops.* Reading, UK, The University of Reading.

Thompson, A.K. 1996. *Postharvest Technology of Fruits and Vegetables.* Blackwell Science, Ltd., Oxford, UK, 410 pp.

Wells, James S., *Plant Propagation Practices.* 1985. American Nurseryman Publishing Company, 111 North Canal Street, Chicago,Illinois 60606.

Part III : Cultivation of Temperate Flowers

18

Chrysanthemum

Chrysanthemums are one of the most popular flowers in the world, next only to the Rose. Chrysanthemum was named by Carolus Linnaeus from two Greek prefixes, 'Chrys', which means golden (the color of the original flowers), and 'anthemon', meaning flower.

Chrysanthemum flowers bloom in various forms, and can be daisy-like, decorative, pompons or buttons. Chrysanthemum blooms come in a huge variety of shapes and sizes. Chrysanthemums come in wide range of colors. In addition to the traditional yellow, other popular colors are white, purple, and red. They are very poupular in floral bouquets and flower arrangements. Gift chrysanthemum flower arrangements to your beloved to tell them how much they mean to you.

Each Chrysanthemum flower head is actually a cluster of many flowers, composed of a central group of short disk flowers surrounded by rings of longer ray flowers. Chrysanthemums are classified into nine categories according to the type and arrangement of disk and ray flowers - Incurved, Reflexed, Intermediate, Late Flowering Anemones, Singles, Pompons, Sprays, Spiders/Spoons/Quills, Charms and Cascades. For example, the 'reflexed' Chrysanthemum consists of ray flowers that curve downward into an umbrella shape; the 'quill' has tubular ray flowers that radiate from the center of the head.

GROWING CHRYSANTHEMUM FLOWERS

Chrysanthemums means 'golden flowers' , 'Chrys', which means golden and 'anthemon', which means flower combine together to give this name to Chrysanthemums. They were named so because of the golden color of its original flower. However, chrysanthemums are more commonly known as 'mums'. It is a genus of approximately 30 species of perennial flowering plants in the family Asteraceae, found in Asia and northeastern Europe. They are one of the most vivacious perennial flowers. They are found in several colors like white, pink, yellow, lavender, bronze, salmon, orange, or red. They also come in various shapes and designs. Petals of chrysanthemum flowers can be either single or in daisy form.

Figure 1. Chrysanthemum Flowers

TYPES OF CHRYSANTHEMUM FLOWERS

There chrysanthemums flowers are of several types. They are broadly categorized into nine categories as per their types and placements of disk and ray flowers. The nine categories of chrysanthemums as per this classification are Incurved, Reflexed, Intermediate, Late Flowering Anemones, Singles, Pompons, Sprays, Spiders/Spoons/Quills, Charms and Cascades. Chrysanthemums are regarded as the flowers of month November.

Tips on Growing Chrysanthemum

- Chrysanthemums can easily be growr and are propagated by dividing the roots, cuttings and seeds.
- They are not even specific to any season. However sunny weather is the most preferable weather for them.
- They are generally grown in the area which observes low rainfall.
- While buying chrysanthemums plants you must take care not to buy plants with fully bloom flowers buds. Such plants have shorter booming period.

- Generally the chrysanthemums plants flower after three months of sowing.
- The chrysanthemums are generally pinched after few days to make them bushy.
- It is preferable to segregate the branches of the chrysanthemums plants from one another and sow them with the stems of perennial flowers, which have healthy roots.
- A well-drained soil is the best kind of soil for chrysanthemums. It is also necessary to add compost to the soil once a year, if you are planning to grow chrysanthemums.
- You should apply one-pound fertilizer per 100 square feet of chrysanthemum bed during their budding phase to help them grow in a better manner. Also, keep in concern that the fertilizer you mix with the soil should be mixed within 6 inches of the soil to give proper nutrition to the chrysanthemums plants.
- Provide them with thorough irrigation so that the roots evenly absorb the fertilizer.
- The chrysanthemums flowers should be sowed about 18 to 24 inches away form one another.
- The faded Chrysanthemum flowers should be removed from the plants on a regular interval as it hampers the growth of other blooms.
- Once these chrysanthemums flowers are six inches long, you must cut short one or two inches from their stem. You must continue this practice till the beginning of August. This is done in order to make them bushy.

Varieties

Ajina Purple

This variety is developed by NBRI, Lucknow. It is a regular incurve type of chrysanthemum with very large flowers. Ray florets of this variety are narrow to broad. Florets are very smoothly incurved in a pattern to form a perfect ball. Disc is not visible.

Snowball

This variety is developed by NBRI, Lucknow. It is a regular incurve type of chrysanthemum with very large flowers. Ray florets of this variety are narrow to broad. Florets are very smoothly incurved in a pattern to form a perfect ball. Disc is not visible.

Potomac

This variety is developed by NBRI, Lucknow. It is a single large flowered type of variety. Ray florets of this variety are long, elongated and straplike. Number of whorl of florets restricted upto four. The disc is conspicuously visible.

M-24

This variety is developed by NBRI, Lucknow. It is a large flowering variety of spoon type. Ray florets of this variety are tubular with spatula like open tips. The size of open portion varies. Disc visible.

Agnishikha

This variety is developed by NBRI, Lucknow. It is a gamma ray induced new mutant. It is a small flowering variety of decorative type producing erythrite red flowers. Ray florets are strap-like. Number of whorls of ray florets is more than five. Disc is not visible due to developed ray florets.

Batik

This variety is developed by NBRI, Lucknow. It is a gamma ray induced new mutant, which produces unique colour combination of yellow stripes on red background.

Harvest Home

This variety is developed by NBRI, Lucknow. It is a small flowered variety of stellate type. Ray florets of this variety are strap-like but both the side of ray florets are reflexed downward. Florets may not be twisted. Disc flat with short florets.

Gypsy Queen

This variety is developed by NBRI, Lucknow. It is a large flowering variety of sipder type. The ray florets are tubular and elongated with tips open or closed. Tips may be open or closed but in either case they are coiled or hooked. The rays may either fall or spread.

Navneet Yellow

This variety is developed by NBRI, Lucknow. It is a gamma ray induced new mutant, which produces yellow flower heads.

Gamit

This variety is developed by NBRI, Lucknow. It is a large flowering variety of reflexed type. Ray florets of this variety are narrow to broad. Ray florets bent backward and downward. Inner florets remain incurved at the early stage concealing the disc florets of the bloom. Outer florets turn outward away from the central tuft. Blooms look globular but may be somewhat flattened. Average bloom size 15-20 cm.

Nanako

This variety is developed by NBRI, Lucknow. It is a small flowered variety of pompon type. Ray florets of this variety are short, broad and very systematically and uniformly arranged to give bloom a contact hemispherical shape. Width and breath almost equal. Ray give florets may be incurved or reflexed. Disc normally covered or inconspicuously open.

Gauri

This variety is developed by NBRI, Lucknow. It is small flowered variety producing whit flowers with yellows heads in the centre. A gamma ray induced new mutant of this variety has also been developed which produces yellow colour flowers.

Rosa

This variety is developed by NBRI, Lucknow. It is a small flowering variety of anemone type. Disc florets of the flowers are well developed and prominent. Ray florets may be flat, twisted and quilled.

Shabnam

This variety is developed by NBRI, Lucknow. It is a small flowering variety of decorative type. The disc is not visible due to developed ray florets. Ray florets regular or irregularly reflexed.

Taruni

This variety is developed by NBRI, Lucknow. It is a large flowered variety of reflex incurve. Ray florets of the flowers are narrow to broad. Ray florets bent backward and downward. Inner florets remain incurved at the early stage concealing the disc florets of the bloom. Outer florets turn outward away from the central tuft. Blooms look globular but may be somewhat flattened. Average bloom size 15-20 cm.

Pournima

This variety is developed by NBRI, Lucknow. It is a large flower pompon type producing white flowers. Ray florets of the flowers are short, broad and very systematically and uniformly arranged to give bloom a contact hemispherical shape. Width and breath almost equal. Ray give florets may be incurved or reflexed. Disc normally covered or inconspicuously open.

Soil & Climate

Soil

The ideal soil for chrysanthemum growing is a well-drained sandy loam of good texture and aeration. The soil should be neutral or slightly acidic with pH 6.5 to 7.0 having high organic content. Very light sandy soils are not recommended owing to their poor moisture holding properties.

Climate

Light: Light is one of the most important climatic factors influencing the growth and flowering. The flowers of chrysanthemum are very much influenced by the quantum and quality of light; most of the cultivars require shorter days for flower bud initiation and development. Under long day condition, chrysanthemum tends to remain vegetative. As it is a short day plant, it naturally flowers in autumn and winter.

Temperature: Based on the temperature requirements chrysanthemum cultivars are classified into 3 groups

1. Thermozero cultivars: Varieties, which flower at any temperature between 10-27°C, but most constantly 16°C night temperature.
2. Thermopositive cultivars: A minimum of 16°C required for initiation and at 27° C there will be rapid initiation but delayed flowering.
3. Thermonegative cultivars: Bud initiation occur at low or high temperature between 10°C and 27°C but continuous high temperature delay bud development.

Altitude: It can be successfully grown upto an altitude of 1200 m elevation.

Relative Humidity: A relative humidity of 70-90% is optimum for growing chrysanthemum.

Propagation

Chrysanthemum is propagated vegetatively either through root suckers or terminal cuttings.

Suckers: Suckers arise from the underground stem and these are separated and planted in prepared nursery beds during January for stock plants. Regular pinching is performed in these plants for vigorous and profuse branching. Some of these stock plants are used for preparation of cuttings. The first pinching is rd performed in April, followed by monthly pinching upto June. After 3 pinching in June, cuttings are taken from these mother plants.

Terminal Cuttings: Cuttings of 5-7 cm in length are taken form healthy stock plants in June. The cuttings are prepared by removing basal leaves and reducing the leaf area of remaining leaves to half. The basal portion (less than half inch) of cuttings are dipped rooting hormone (1000 ppm solution of IBA.) for better rooting. Sometimes the lower portion of cuttings is treated with some copper fungicide to avoid fungal growth. These rooted cuttings are ready for planting in the field.

INTERCULTURAL OPERATIONS

Weed Control

Weeding and hoeing are yield generally done manually as and when required, normally 8-10 times yearly. Crop suffers heavily if timely weeding is not given. Besides, control of weeds the soil is made loose porous to provide aeration.

Pinching

Pinching is one of the most important operations in chrysanthemum culture. Pinching refers to the removal of the growing tips of the plant to induce the growth of vegetative laterals. It reduces the plant height, promotes axillary branching, delays flowering and helps in breaking rosetting. Pinching is performed both in suckers and in cuttings. It is normally done with thumb and forefinger. Pinching is most essential for small flowered chrysanthemum. First pinching is done when the plants reach a height of 15-20 cm with 3-4 pairs of leaves. A second pinching may be necessary if the plants make straggly and lean growth. Two types of pinching are performed in chrysanthemum. In soft pinching the soft tip of the shoot along with 2-3 open leaves is removed while in hard pinching a longer portion upto hard shoot is removed. In case of standard chrysanthemum only single bloom on a branch is usually allowed to produce.

The pinching is not done if only one central bloom is desired on the main branch. Single pinching is done, if two flowers are desired, whereas double pinching is done for four flowers. In spray chrysanthemum numerous small to medium sized flowers are produced, therefore, two pinchings are required to encourage lateral growth. As a general rule rooted cuttings are pinched 2 weeks after planting or approximately 100 days before full bloom.

Disbudding

This operation is mostly performed for large flowered and decorative type chrysanthemum. Disbudding method vary according to the type of chrysanthemum grown. Many of the varieties are disbud or standard types, in which the largest terminal bud is retained and all axillary buds are removed. Disbudding of spray varieties is very easy because in this case only the large apical bud is removed and the axillary buds are allowed to develop. There is no specific rule for disbudding of spray varieties; it varies with the type of spray produced. When growers want to develop three blooms per plant or one bloom per plant these operations are most essential. Disbudding operations is an important factor in the maintenance of high-quality product

De- suckering

During the vegetative growth phase, plants grow upwards. New suckers continue to develop from base of plants. For proper and vigorous growth of plants, suckers are removed from time to time.

Irrigation

Chrysanthemum requires frequent and thorough irrigation. If rains are delayed irrigation is given twice a week in the first month and subsequently at weekly intervals.

Plant Growth Regulators

Crop growth regulation and flowering can be modified or controlled by use of growth regulators. Flower quality and yield can be improved by the use of regulators. Spraying GA (50ppm) at 30, 45 and 60 3 days after planting increases the flower yield.

Manuring & Fertilization

Chrysanthemum is a heavy feeder and requires large amount of both Nitrogen and Potassium. Nitrogen is required at early stage and the plants need P throughout the growth period. As the buds appear, the proportion of K should be increased and N should be reduced.

CHRYSANTHEMUM DISEASES

Wilt

Initial symptoms are in the form of yellowing and browning of leaves. Affected leaves die from the base of the plant upward. Infected plants are stunted and often fail to produce flower. Wilting may cause rotting of root or the base of the stem.

Control: Since the disease is a soil borne it is difficult to control with chemicals alone. Drenching the soil with Dexon or Carbendazim is very effective. Before planting dipping the rooted cuttings in a solution of Thiram @1.5g/litre of water. Since the disease spreads mostly through cuttings, it is important to use disease free planting material. Disease can further be minimised by following strict sanitation; periodical monitoring; crop rotation and roughing of infected plants.

Rust

Rust is a serious disease especially in the early spring. The disease symptoms are in the form of brown blister-like swellings, which appear on the undersides of leaves. These burst open releasing masses of brown, powdery spores. Severely infected plants become very weak and fail to bloom properly.

Control: Early removal of infected leaves/plants helps to prevent the further spread of the disease. Spraying the plants with Karathane @0.025% or dusting with Wettable Sulphur @0.2% is effective in controlling the disease.

Septoria Leaf Spot

Leaf spots occur during cool-wet periods of the rainy season. Since the pathogens are spread through rain splashes the lowermost leaves get infected first. Serious infection may result in premature withering of the leaves; the dead leaves hang to the stem for some time. When flowering starts the infection occurs on flower buds, which rot completely.

Control: These disease can be controlled by spraying Bavistin @0.05% six times at 15 days intervals from the end of July or spraying Benomyl (0.1%) followed by Captafol (0.2%) and Carbendazim (0.1%). Destruction of disease debris and avoiding excessive irrigation is recommended.

Stem Rot

The disease is noticed in cuttings, stock plants and flowering plants. It is serious during rainy weather. The affected plants show different symptoms such as leaf chlorosis, necrosis, decay and discoloration of internal portion of the stem. When the flower buds are about to open, small dark streaks are seen at the base of the stem. Root decay is noticed only in advanced stages of infection. In many cases the infection is devoid of any external symptoms and roots also look healthy. In some cases stem and root rot are confined to one side of the stem, the other side remaining healthy.

Control: Cuttings should be taken from healthy plants. Following strict field sanitation, roughing of infected plants and crop rotation helps to minimize the disease incidence. Spraying with Bavistin (0.1%) or Bordeaux mixture (1%) and avoiding excessive moisture in beds help in checking the disease.

Powdery Mildew

Infection is more severe in older plants under humid conditions. The growth of the fungus on the leaves appears as powdery coating. Infected leaves turn yellow and dry out. Infected plants remains stunted and fail to flower.

Control: Disease can be effectively controlled with Sulphur fungicides or Captan (0.2%). Good ventilation and proper spacing for free circulation of air is recommended.

CHRYSANTHEMUM HARVESTING

The correct stage of harvesting depends upon the cultivar, marketing and purposes etc. The flowers are harvested 3 months from the date of planting at an interval of 4 days. For floral decoration and garlands, fully opened flowers are plucked with or without the peduncle. Flowers are packed loose in bamboo baskets or gunny bags and sent to retail markets for sale.

Generally harvesting is done early in the morning. The stems are cut 20 cm from the ground and are collected in trays or buckets containing cold water (15-18°C) and a floral preservative (Silver nitrate 25 ppm). Alternately, the harvested stems are kept in 1000 ppm Silver nitrate, for 10 seconds-10 minutes followed by placing in deionized water to prevent the blockage of xylem by the microorganisms.

The yield of flowers varies according to types of cultivars, growing region, plant density and other management practices. The average yield of loose flowers varies widely from 8-15 tonnes/ ha. In southern states, the crop remains in the field for about ten months if it is ratooned (six months for plant crop an four months for ratoon crop). Ratooning is seldom done now days.

REFERENCES

Carvalho, S. M. P., H. et.al. 2005. "Temperature affects chrysanthemum flower characteristics differently during three phases of the cultivation period". *Journal of Horticultural Science and Biotechnology* 80(2): 209-216.

History of the Chrysanthemum. National Chrysanthemum Society USA

Van Der Ploeg A. and E. Heuvelink. 2006. "The influence of temperature on growth and development of chrysanthemum cultivars: a review". *Journal of Horticultural Science and Biotechnology* 81(2): 174-182.

19

Flowers and Foliage from Plants of Protea Family

The demand for flowers and foliage from plants of the protea family for export has resulted in a significant increase in the number of properties producing this crop. Careful selection of planting material and correct management are essential to ensure a top quality product. The international market for cut flowers and foliage is discerning and highly competitive. Growers should accordingly plan their plantings to achieve a high quality product to coincide with periods of demand. Varieties selected should be those with flowers and foliage that are acceptable to the end user and provide an economic return to the grower.

To determine the suitability of a species, variety, or clone for export, intending growers should consult leading growers and exporters who have experience with Proteaceae. Nurserymen can help with plant selection but will often be swayed by what they have available to sell. Some "crystal ball" gazing based on market feedback is also warranted, as the time lag from planting to full production is often more than three years. Because of this time lag, there is a requirement for alternative income sources in the first few years, and many growers do supplement their incomes with annual crops. It is important to realise that similar-looking varieties can often have very different handling and storage properties.

VARIETIES

Proteaceae known to be currently trading successfully from New Zealand to international markets are listed below. The selection is based on available information. Persons making selections for commercial plantings should base their final choice on whether that particular variety when grown in their locality will produce flowers or foliage that:

— Have the capacity to travel.

— Possess good vaselife, colour, texture, stem length, and presentation.

- Are available at high demand periods.
- Are a viable profitable investment.

Protea cynaroides (King): Well known and established wherever proteas are sold. Stock planted should be of variants chosen for long stems, good flower form and colour, and appropriate flowering time.

Protea magnifica (Queen): Well established on all protea markets. Stock used should be of a variant known to produce well in the intended area. Does not grow well in all areas.

Protea neriifolia, P. compacta, P. eximia: Established on markets. Selected clones which are not susceptible to leaf blackening should be grown.

Leucospermum cordifolium: Select variants which have good colour (bright orange), stems 45 cm or longer, clean foliage (resistance to fungal attack), and an appropriate flowering time.

Leucospermum conocarpodendron: Bright yellow flowering species.

Leucadendron: Female forms are generally preferred, except for L. discolor and certain of its hybrids, which are male.

PROPAGATION AND CULTIVATION

Most forms of South African and Australian Proteaceae grown for cut flowers can be propagated vegetatively as well as by seed. While seed has been the traditional method, vegetative propagation is the only technique by which selected forms suitable to the trade can be maintained. Seed propagation is now only widely used to propagate banksias.

Natural Habitats

Basic requirements for the successful cultivation of proteaceous plants (based on natural habitats) include:

- An open airy situation in full sun.
- An acid soil that is preferably fairly light in texture and free-draining. Heavier soils that are not well drained can be suitable if they slope to allow excess water run-off.
- Soils that are not excessively high in available nutrients, particularly phosphorous.
- Situations not subject to heavy frosts.

The soils in which proteaceous plants are found growing naturally are usually very acid and low in nutrients. The plants have evolved a complex root structure known as the "proteoid" root system to enable them to extract both nutrients and moisture under harsh conditions. Keeping these delicate, waxen, white rootlets alive and active is often the source of problems associated with successful cultivation of Proteaceae.

Figure 1. Flower of King Protea

Soil Type and Nutrient Requirements

A free-draining soil is essential. Without adequate aeration the proteoid rootlets will often die, and a soil environment suitable for soil-borne fungus diseases such as Phytophthora occurs. The incidence of such disease is lower in light, sandy, very free-draining soils.

When soil tests are carried out it is necessary to look for low nutrient levels. High levels of both calcium and phosphorus can destroy the root system. As a result, although a plant may appear healthy, a period of stress, especially that caused by drought, may result in its collapse. Sometimes this has been unjustly blamed on Phytophthora, as some of the symptoms may be similar.

In the field it is wise to avoid the application of fertilisers completely. Research is being done on the nutrient requirements for proteaceous plants and although nutrients should be incorporated in soil-less container mixes, it is still considered risky in the open ground situation.

Practical experience has demonstrated that the following soil testing figures can be used as a working guide (based on MAF quick soil tests; QT values followed by ppm in soil):

- pH - a maximum of 6.5, with an ideal of 5 - 5.5. (In their natural habitats, some species such as P. cynaroides may occur in soils with a pH of 3.5.)
- Calcium - a level of less than 10 (< 1250 ppm in soil) is desirable.
- Potassium - less critical, but under 20 (< 400 ppm in soil) is preferable.
- Phosphorus - an upper limit of 30 (33 ppm in soil) may be permissible, but ideally should be under 20 (22 ppm) (except for 'Safari Sunset', where the upper level may be as high as 45 (50 ppm)).
- Magnesium - under 30 (<150 ppm in soil) is preferred.

A significant finding by South African researchers is that genetic variation within species may affect phosphorus tolerance. In problem soils, clones may be selected that exhibit sufficient tolerance to enable successful cultivation.

Irrigation

Owing to the complex root system, irrigation should not be necessary once plants are established, unless there is an extended period of drought.

Irrigation could be needed in situations where the proteoid root system does not develop (e.g. owing to high nutrients) or on light soils subject to summer drought (e.g. coastal sands).

Air Circulation and Shelter Requirements

Proteaceous plants generally dislike sheltered humid environments. They prefer good air circulation around the bushes, as this assists the drying of the foliage and reduces the risk of fungal infection.

Only a few species will tolerate gale-force winds as most have comparatively shallow root systems, and if these are disturbed through wind movement the resultant root damage may allow pathogens to enter the plant. This type of damage due to wind is more serious than foliage burning, which is not often a problem. In exposed wind-prone situations shelter belts which create a "filter" effect are recommended, as opposed to dense heavy shelter. Fairly wide-spaced trees such as Casuarina are preferred.

Planting Distances

These are determined by the ultimate size of the bushes and the method of property maintenance. For example, if a tractor of 1m width is used for spraying and mowing and the species is *Leucospermum cordifolium*, which has a normal spread of about 2m, then the rows will need to be a minimum of 3m apart (centre to centre).

Planting *Leucadendron* 'Safari Sunset' as close as 60 cm apart in an effort to achieve longer stems is not recommended. Cramming can result in premature losses through root competition and disease.

Penetration of sprays is more difficult in densely planted rows. In summary, some air space between bushes is regarded as desirable to give longer life to the bushes and fewer maintenance problems.

Planting Techniques

Compacted ground should be sub-soiled to break up any pan that could impede drainage.

Where annual weeds and grasses are a problem, spray a 0.6-1 m strip with a desiccant herbicide. Problem perennial weeds such as couch, paspalum, and similar should be sprayed with a suitable non-residual herbicide. If the weed problem is widespread, the entire block should be sprayed off.

The strips where the shrubs are to be planted should ideally be rotary-hoed. This enables the roots to spread out quickly into the surrounding soil. When planting out into rotary-hoed ground, the bushes should be planted slightly deeper than appears necessary to allow for the soil to sink and compact. Digging holes in straight pasture often slows the growth of the plant, because the roots have to penetrate the harder surrounding ground.

Support

In some commercial plantings post and wire support is used. This method is permanent and cheaper than staking. It involves clipping the main leaders of shrubs to wire, using PVC ties of the type used for kiwifruit. The end posts are blocked, using timber or boulders to reduce movement when the wire is tensioned. Use 10 gauge galvanised wire and in-line permanent strainers. This system also provides a straight line for planting. The height of the first wire will differ according to the size of the bush being planted, but is usually about 15 cm from the ground. A second wire is installed as the bushes grow and is normally about half-way up the posts. As the bushes grow and the second wire is utilised, the PVC clips may need to be removed from the first wire to prevent ring-barking.

Pruning

Pruning is a controversial subject.

Leucadendrons need to be well cut back at the end of the flowering season to encourage plentiful new clean growth. On vigorous hybrids such as 'Safari Sunset', it may also be desirable to thin some shoots to encourage longer stems. Waratahs need to be pruned well down after flowering to encourage vigorous new flowering stems that can be picked easily. Proteas only need to be shaped, with straggly branches eliminated and unpicked flowers cut back to the point just above the new season's growth. Leucospermums and banksias need little pruning.

Weed Control Methods

Mulching

Although weeds can be controlled by mulching, caution should be exercised in the choice of material.

Porous "weed matting" materials are now the norm for weed control, with the width used and method of laying dependent on crop and site.

Wide black polyethylene mulches can reduce aeration and may have long-term detrimental effects.

Sawdust and straw mulches tend to take nitrogen from the soil during their decomposition process, which could be a disadvantage. Such mulches also tend to cause types of collar rot - particularly if sawdust builds up around the plant in windy situations. Pathogens may also be harboured in such mulches.

Herbicides

The use of herbicides is the most common form of weed control. Post-emergence herbicides can be used for emerged weeds, and several residual herbicides show promise for longer term weed control. It is important to minimise any contact between the spray and the crop by the use of low pressure sprays and anti-drift guards.

Hand hoeing and "dust mulches"

This method is still widely used. As many proteaceous plants have roots near the surface it is very important that hoeing is shallow.

Mowing inter-row strips

Mowing is the most common method of inter-row grass control. Avoid clippings being thrown into foliage or around the trunks of plants. They can cause blemishes on the foliage and collar rot around the trunks.

In more intensive plantings, where it is not possible to manoeuvre machinery between rows, either hand-hoeing or chemical weed control is normally used.

Diseases and Pests

In their native habitat Proteaceae are subject to a wide range of pests and diseases.

The plants grown in NZ are in general less subject to attack by diseases and pests than many others. However, commercial cut flower production, particularly for export, requires a clean healthy blemish-free product, therefore plants should be protected from pests and diseases.

Root diseases

These include:

— *Phytophthora root rot (Phytophthora cinnamoni):* Phytophthora root rot (PRR) can infect most Proteaceae grown in New Zealand. The fungus occurs naturally in all soils, particularly those that are poorly drained.

PRR is usually first observed as wilting and collapse of plants. Infected roots show browning and rotting. 'Wiriwiri wilt' of *Leucadendron* spp. is now thought to be due to PRR infection.

The best means of control is prevention. Ensure that any proposed sites are adequately drained and that surface ponding does not occur following heavy rain. Where the disease occurs, obviously infected plants should be removed and destroyed. Other plants in the vicinity should have a soil drench of metalaxyl ('Ridomil MZ'), 'Terrazole', or fosetyl-Al ('Aliette'), applied around the root zone. Ensure the soil is thoroughly soaked around each plant to ensure the chemical reaches the roots. Alternatively, foliar sprays of a potassium phosphite-based compound (e.g. 'Foliafos') can be used. Be aware that resistance can occur, so alternate the compounds used. While these chemicals can control PRR, no official recommendations on specific use have been made for Proteaceous crops, and you should check with your chemical supply representatives for latest advice.

— Other soil fungi of less significance occasionally found associated with root rots of proteaceous plants are *Cylindrocladium* and *Fusarium* species. These fungi are generally only a problem when induced by poor drainage.

Leaf and stem diseases

These include:

— *Grey mould (Botrytis cinerea):* Recorded in NZ and overseas, causing stem lesions and infecting flowers of Leucospermum species. Usually seen as a grey-coloured fungus on the infected areas.

Infection can be reduced by ensuring good air circulation within the crop. Where the disease is a problem, regular sprays of vinclozolin ('Ronilan') should be applied.

— *Wilt (Verticillium dahliae):* This soil-borne fungus has been recorded in NZ on Protea compacta. Verticillium infects the vascular system of plants, causing wilting and dieback. The recommended method of control is to remove the infected plants.

— *Silverleaf (Chondrostereum purpureum):* Recorded on Protea species and on Leucadendron 'Safari Sunset' and 'Red Gem' in NZ, causing silvering and, in severe cases, yellowing of foliage on affected stems. Entire plants may eventually become infected.

The silverleaf fungus enters through pruning wounds or injury sites. To help prevent infection, pruning should be carried out during dry weather and a fungicidal spray such as captafol ('Difolatan') applied to pruning cuts. Only major pruning cuts or injury are considered to be at risk, not normal cuts associated with the removal of flower stems.

— *Bacterial leaf spot (Pseudomonas syringae):* Recently identified as causing a leaf spot on Protea cynaroides, where it is usually observed as small (up to 5 mm diameter), occasionally sunken, dark-brown or black well defined spots.

Bacteriocides used for control of *P. syringae* in fruit trees could be tried where the incidence of the disease is of concern.

Other diseases

The causes of the following diseases are unknown:

— 'Waitara wilt': The symptoms of 'Waitara wilt' occur on Leucadendron species in a number of areas. It is usually first seen as a watersoaked area on a leaf which rapidly enlarges and dries out to form a pale tan necrotic area, often encompassing the entire leaf. In severe cases some stems of plants or even entire plants may be affected, with little or no live leaf tissue remaining. Some plants affected in this manner may apparently recover for a period, but then succumb to the problem again at a later stage.

— Bract discolouration: Leucadendron laureolum and L. microcephalum show a rusty discolouration of the terminal leaf bracts.

Pests

— *Leaf roller caterpillars (Ctenopseutis obliquana; Epiphyas postvittana):* Regarded as the most significant insect pests of proteaceous plants. The typical symptoms are holes in the leaves or bracts, or chewing damage around the margins. The formation of webbing on leaves or leaves webbed together is also indicative of these pests.

Applications of insecticides such as acephate ('Orthene') are recommended at regular intervals to prevent attack.

— *Scale insects (Phenacaspis eugeniae - 'waratah scale'; Parlatoria pittospori; Coccus longulus):* These are white or brown limpet-like insects, usually found on the underside of leaves.

Severe infestations result in yellowing or browning of the upper leaf surface - corresponding to the point of attachment of the insect on the underside, and stunting of plant growth can result.

A regular spray programme of insecticide such as maldison ('Malathion'), plus an all-seasons oil, is normally recommended for control. Orthene is also quite useful.

— *Mites (Tetranychus urticae - two-spotted mite; Panonychus ulmi - European red mite):* Mites cause silvering and mottling of host leaves or flowers, rendering them unsuitable for export.

A regular spray programme using a miticide such as cyhexatin ('Plictran') is recommended, particularly during hot dry periods of the year.

— Other pests of less significance recorded on proteaceous plants include aphids, cicadas, leaf miners, mealy bugs, and psyllids, all of which should be controlled by regular preventative insecticide application.

Spray Programme

As a general guide the following spray programme for disease and pest control is suggested:

— benomyl ('Benlate'), plus permethrin and pirimiphos-methyl ('Attack'), plus cyhexatin ('Plictran'). Take care with the use of benomyl as there is a tendency for resistance to develop.

Alternate with:

— chlorothalonil ('Bravo'), plus acephate ('Orthene').

A protective type of fungicide such as 'Mancozeb' and a variety of potent fungicides can be used whenever fungal problems occur. This practice can be more economic as some of the potent fungicides are too costly to be used on a regular basis.

Applications can be made at monthly intervals after harvesting has been completed, reducing to 10-14 day intervals during periods of active growth, and to 7-10 day intervals approaching harvest.

These are not the only chemicals that may be used and alternatives are available.

There is no assurance that flowers and foliage treated with chemicals in the above programme will be acceptable to all countries to which they may be exported.

Frequent inspections should be made to check for the presence of pests and diseases. With experience, the interval between sprays and the types of sprays used may be varied to suit individual situations. However, once damage occurs it cannot be reversed and can adversely affect the profitability of the crop.

Harvesting

It is important to pick at the right stage of development to ensure that blooms travel well and have a good vase life on arrival.

Harvesting should be completed in cool conditions if possible, either early in the morning or late in the afternoon when carbohydrate reserves in the leaves could be expected to be at their highest.

Maturity at harvest determines both vase life and product quality. Less mature flowers have a longer vase life, as long as they are not picked too immature. They may also place a heavier demand on the carbohydrate reserves of the leaves in blackening-susceptible cultivars.

If flowers can be picked as the first florets are splitting, this will reduce the problem of nectar being released into the package and spoiling the product.

Leucadendron

Most of the popular picking varieties are female forms, and the main criterion is to pick when the bracts (the terminal leaves which surround the central seed-bearing cone) are at their most colourful and yet remain firm.

Male flowers must be harvested when the crop is showing good colour, but before the pollen is shed. Experience will help identify these stages.

Leucospermum and Banksia

Leucospermums and banksias should be picked just as the first styles begin to unfold. This results in less damage in transit, more stems per carton, and maximum vase life at the other end.

Protea

Proteas should be picked just as the flower begins to open. If the bud is too tight when picked it will not usually open. Flowers that are too far open have a shorter vase life and are bulkier to pack.

Telopea (waratah)

Generally telopeas should be picked just as the first styles start to open. Stem length is no problem as stems are generally very long.

Stem length

Maximum stem length is needed for export. Leucadendrons, leucospermums, and banksias can be picked almost as far down into the bush as possible without affecting future growth. Protea stems generally should not be cut below the previous season's growth as they do not generate much new wood from old growth. P. cynaroides is an exception and benefits from being cut hard.

Bunching and tying

Most larger-flowered varieties are not bunched. Slimmer-stemmed leucadendrons are usually bunched into 5s or 10s, but this depends on market requirements.

Postharvest Handling

The newly cut flowers should be placed in tap water (room temperature) and transferred to a cool room (approximately 4°C) for 3-5 hours or overnight. Good quality water should be used for holding the stems. If the water supply is high in salts then rain water should be used.

Tests with floral preservatives have given inconsistent results and holding in water appears to be the most satisfactory treatment at this stage. However 50 ppm chlorine (e.g. 1 teaspoon of Janola in 5 litres of water) is a useful precaution against bacterial growth.

Dependent on the market destination flowers may need to be treated with Floragas (disinfestation treatment) before or after packaging.

Flowers should be packed to restrict moisture loss. Waxed paper lining the carton or flower sleeves can be used. Exact packaging requirements need to be discussed with individual exporters, as often their requirements differ.

The packed flowers should be recooled using forced air cooling and kept cool throughout the transport period. It should be emphasised that cooling and refrigeration during transit are very important for immature flowers to enable them to open. Some exporters require icepacks to be used within the cartons to maintain cool temperatures during transit.

Leucadendron

The flowers of the red leucadendrons (cultivars 'Safari Sunset' and 'Red Gem') do not bruise as easily as the other Proteaceae and are long-lived in the vase. They can be stored at 0 - 2°C for 2 - 3 months, though vaselife decreases with increased length of storage.

Decay in storage can be a problem. Stems to be stored should not have any foliage stripped from them, and fungicides such as 'Benlate', 'Rovral', and 'Ronilan', which are active against Botrytis spp., should be used as a pre-storage dip.

Other types of leucadendrons do not store as well as 'Safari Sunset' and 'Red Gem' and should therefore be cooled rapidly after harvest and handled quickly.

Holding in water during cooling and rapid transport to markets is necessary for most leucadendron material.

Leucospermum

Maturity of leucospermum flowers at harvest determines their vase life. Over-mature flowers are generally shorter lived, but immature flowers do not open well in the vase.

The flower heads are easily broken from the stems, and problems can occur in packing with interlocking of styles from adjacent flowers. This results in damage when the flowers are unpacked, and some method of keeping the heads separate in the pack would be worthwhile, e.g. head sleeves.

Crushing of the styles and release of nectar are problems with more mature flowers.

Protea

Protea flowers harvested as loose buds open satisfactorily but tend to have poorer colour than more mature flowers.

Holding in water and rapid cooling of flowers are beneficial postharvest treatments.

REFERENCES

Pirone. P. O. 1978. *Diseases and Pests of Ornamental Plants*, 5th Edition. John Wiley and Sons. 605 Third Ave. New York, NY 10158.

Society of American Florists. 1985. *Care and Handling of Flowers and Plants.* 1601 Duke Street. Alexandria, VA 22314 200 pp. $35.00

Stevens, Alan. 1996. *Field Grown Cut Flowers: A Practical Guide & Source book.* Avatar's World Ph. 800.884.4730

20

Rose

Roses are the most popular flowers in the world and have probably always been the most popular flowers. Roses have been symbols of love, beauty, war and politics from the beginning of time. Different varieties and colors of Roses denote different Meanings. Rose is popularly known as the flower of love, particularly Red Roses.

Figure 1. Rose flower

Rose, besides being the popular gifts around the world, are also a great addition for the homes and offices as decorations. A rose flower in a drinking glass arranged on a table to a floral arrangement of a bunch of roses add that extra touch to the decor. Besides fresh cut roses, artificial flowers like silk roses in different colors are also widely used. Order your silk rose now and add grace to your decoration.

The birthplace of the cultivated Rose was probably Northern Persia, on the Caspian, or Faristan on the Gulf of Persia. Historically, the oldest Rose fossils have been found in Colorado, dating back to more than 35 million years ago. Roses were considered the most sacred flowers in ancient Egypt and were used as offerings for the Goddess Isis. Roses have also been found in Egyptian tombs, where they were formed into funerary wreaths.

Confucius, 551 BC to 479 BC, reported that the Imperial Chinese library had many books on Roses. Ancient Sumerians of Mesopotamia mentioned Roses in a cuneiform tablet written in approximately 2860 BC. The English were already cultivating and hybridising Roses in the 15th Century when the English War of Roses took place. The winner of the war, Tudor Henry VII, created the Rose of England by crossbreeding other Roses. While no Black Rose yet exists, there are some of such a deep Red color as to suggest Black. Roses are omnipresent and grown over all parts of the globe.

The Netherlands, with about 8000 hectares of land under Rose cultivation, is the global leader in Rose cultivation. 54 per cent (about 5000 hectares) of the cultivated land in Ecuador is under Rose cultivation! Zambia, a small nation, had 80 per cent of its cultivated land under Roses.

CLASSIFICATION OF ROSES

There are many types of roses available in garden centres, so it's important to choose the correct one for your particular situation. Luckily enough, all garden centres use the same names to describe the different types. All the common types of rose are described below.

Wild Roses

These are the parent roses of all other roses, found growing in the wild. Most have small single flowers with five petals. The colours are normally pink, rose, yellow or white rose. Wild roses are very hardy and grow as largish bushes or climbers. They are very resistant to disease and pests. The flowers appear only briefly in early summer, often followed by attractive red hips in autumn

Common varieties include *Rosa rugosa* which has largish rose coloured flowers and beautiful red hips in autumn, *Rosa arvensis* with small yellow flowers and Rosa canina with scented pink flowers.

Old Garden Roses

These are hybrid roses which have been bred by crossing different species of wild roses. They make good-sized shrubs and a few produce flowers in two or more flushes throughout the summer, although most only produce flowers in early summer for three or four weeks. Normally they are over 100 years old.

Miniature Roses

With varying heights up to 45cm (18in), the miniature rose is an ideal plant for pots, rockeries and the front of the border. They produce flowers in flushes throughout the summer - the flowers are miniature rose versions of either the cluster-flowered or large flowered roses.

Good examples include 'Angela Rippon', 'Queen Mother', 'Happy Birthday' and 'Snow Magic'. Like most roses miniature rose prefer a full sun position which helps to produce more and brighter flowers. A semi-shade position is OK if they get at least half the day's sun.

Because of their more delicate root structure, miniature roses need a good supply of water at all times. Where they are in pots, this even more important. All miniature roses in open ground will appreciate an annual mulch of well-rotted compost around but not touching the stems. This will retain moisture and provide a slow release source of food.

A regular application of rose fertiliser at the rate described on the pack will benefit them greatly. Stop feeding around for weeks before the last frost date in your area. This will make the plants "harder" and more able to withstand the winter cold.

Floribunda Roses

Floribunda roses were produced by crossing cluster-flowering Polyantha roses and Hybrid Tea roses. They first became popular in the 1930s and continue today with new varieties being produced regularly.The individual flowers on floribunda roses open at different times and this enables them to give a show of colour all the flowering season from mid June to mid September.

Hybrid Roses

The name 'hybrid tea' came about because they originated from the old Tea roses. These Tea roses had a scent of freshly picked tea and produced flowers of beautiful form. However, they were not fully hardy and [illegible] to diseases. After many years of breeding however, Hybrid Teas are now fully hardy and much better at resisting disease.

Hybrid teas are by far the most popular form of rose and there are a huge number of varieties in all colours, perfumes and shapes. They have large flowers and they form small to medium size bushes. They are well suited to being trained as standards.

For the amateur or those with a small garden, these are the ideal choice. They produce a large number of flowers for the space occupied and are the easiest to grow. There are new varieties produced each year, and you are really spoilt for choice. Any of the varieties featured on our pages would be an excellent choice.

Ground Cover Roses

Ground cover roses come in a variety of forms but they share the characteristics of long spreading branches which spread over the ground rather than growing upright.Most ground cover roses are repeat flowering and most have little or no scent. Most of the growth occurs from the base of the plants.

Ground cover roses have thorns which makes weeding between them difficult. They will not exclude all light and loose their leaves in winter. So it's a good idea to apply a thick layer of mulch before planting and repeat every year or so.They require the same treatment as other roses. Pruning consists of keeping the plant to shape and occasionally pruning the centre of the plant to let air circulate.

Shrub Roses

Shrub roses are similar to the Old Garden Roses, but they have been bred over the last 100 years. In general they are more disease resistant and generally flower for longer than the Old Garden Roses.They tend to include roses that are not included in other categories. They are characterised by their repeat flowering and longish stems.

Flowers are most frequently in clusters of single or double blooms. The plant itself is in the same range as floribunda roses, from 1m (3ft) to 3m (10ft) in height and spread.Originally, old shrub roses produced flowers for only a month or so. New varieties now produce repeated blooms from mid June to mid September.The medium sized shrub roses look excellent at the back of a border. The larger ones are often used as specimen plants or hedges.

Shrub roses should be treated the same as other roses. Because they are so good at producing flowers all year long, it is especially important to dead head them. This will ensure a mass of flowers throughout the summer

Rambling Roses

Rambling roses require very similar treatment to climbing roses although pruning can require special treatment depending on the type of rambler. Climbing roses tend to flower repeatedly throughout the summer whereas rambling roses tend to flower once a year. Rambling roses are also best suited to growing through trees or as ground cover. They are not so suitable for growing up vertical structure such as walls and trellis. They grow at great speed and easily cover up eyesores. After flowering they produce some of the most beautiful hips of all the roses.

Patio Roses

Patio roses are simply small rose. Typically they grow to between 45cm (18in) and 60cm

(2ft). They are ideal for pots and tubs, also at the front of a border when grouped in odd numbers. Miniature roses are even smaller, ranging from 25cm (10in) to 30cm (1ft). Patio roses are often used an alternative to, or combined with summer bedding plants. The roses will flower all summer long providing a great display of colour. They are of course totally hardy and being perennials will appear year after year. We think they are very good value.

Patio roses require the normal care that applies to other roses. Their pruning requirements are however minimal and very simple to do. A common mistake is to try to grow patio roses indoors. In most cases this is doomed to failure. The roses are fully hardy and need conditions that are rarely found in a heated house.

Climbing Roses

Climbing roses have been bred to grow up walls and trellis. Typically they have small clusters of medium to large sized flowers. Normally flowers appear in two flushes with a few in between but some climbers produce only one flush.

Normal rose feeding rules apply to climbing roses. Dead-heading the flowers after the first flush will encourage a good second display. But the sheer height of some climbers will make dead-heading difficult. Good support is critical because climbing roses are weighty plants when in full bloom. Check the support each year.

Often, other types of rose have mutated and become climbers as well. Iceberg is a good example. The climbing version has exactly the same flowers and leaf colour but for some reason they have changed into climbers. Check the label carefully when buying.

PLANTING ROSES

Your roses can be bought either in containers or bare-rooted. Bare-rooted roses are sometimes sold by mail order nurseries - the lack of soil keeps down the weight and therefore the cost. Unlike containerised roses, which can be planted throughout the year, it is best to buy and plant bare-rooted rose either in mid April or mid September.

Roses prefer a site in full sun, but will be successful in partial shade, especially if the shade is during the afternoon. The best soil will be free-draining, slightly acid and medium in texture - neither sandy nor clayey. If your soil is not ideal, improve it by digging in lots of organic material. Aim for the best soil, but remember that roses are tolerant plants. The best planting time is mid April or midSeptember. Don't plant roses where other roses have grown in the last three years - this will greatly increase the risk of 'Rose Sickness'.

Prepare the soil by digging it well to at least a spade and half's depth - more if you have the energy. During digging, sprinkle the soil with a good amount of rose fertiliser or general purpose fertiliser. Leave the soil for four weeks before planting in order to let it consolidate.

Bare-rooted roses need to have their roots spread out and then placed in the planting hole with the bud union 2.5 cm (1in) below the surface of the soil.

Fill in with soil and gently firm it down. Water well if the conditions are dry. If frosts occur soon after planting, it may be necessary to firm the soil down again.

The planting method for climbers and rambling roses is as described above, with a few points to note.

The planting hole should be about 40cm (15in) away from the wall or fence in order to let rain get to the roots. When planting against a wall, be sure to secure the supports against the wall before planting. Finally, water well throughout dry periods for the first year or so.

PRUNING

Roses are very tolerant plants, and with a little care will flourish for many years. If a rose is not pruned, it will turn into a mass of live and dead stems, preventing the circulation of air and cutting out light. Pruning removes old wood, thus allowing air and light to reach all parts of the plant - this helps prevent disease.

Pruning Rambler Roses

At planting time, cut the stems back to around 80 cm (2ft 6in) - this may well have been done at the nursery. Subsequent pruning depends on which type of rambling rose you have.

Type 1

These are the very strong growing rambling roses which can grow as much as 5m (16ft) in a single year. This type only requires pruning to keep it in bounds. Prune as much or as little as you require.

Type 2

This type produces a mass of new shoots mainly from the base of the plant. Flowers are produced from stems that have grown the previous year. After flowering cut back roughly half of the shoots that have produced flowers right to the base of the plant. The non-flowering shoots should be tied to supports so that they are nearly horizontal or fanned out. It is also possible to cut back all the shoots if a rambling rose has got out of control.

Type 3

This type produces most new shoots half way up older branches. Flowers are produced from stems that have grown the previous year. Cut back stems that have flowered back to a new shoot.

Pruning Climbers

At planting time, do not prune at all. If you do, the roses may revert back to bush form in some cases. Instead, fan out the branches so that they are not all pointing upwards. Climbing roses produce flowers more evenly if the branches are fanned out. Each main branch needs to be attached to the support (trellis or wires) with ties.

Annually (in March), little pruning is required other than to keep the plant in bounds and remove any weak or diseased stems. Prune side shoots to about 10cm (4in) long. Older branches should be thinned out by cutting some of them back to about a half of their length. Cut back to just above a new shoot.

Pruning Miniatures

Only prune to cut out dead or diseased wood, and to keep the plant in a good shape. Do not prune on planting. Pruning should cease about 6 weeks before the first frost date in your area.

CARING FOR YOUR ROSES

Roses are not fussy plants, but will provide bigger and more flowers if looked after. The key tasks are feeding, dead-heading, weeding and mulching.

What To Do	*When*	*Comments*
Prune	mid March	early April in cold areas.
Feed	mid March	mid April in cold areas.
Mulch	mid May	early June in cold areas.
Dead-head	July to end of flowering	Do not dead head if you want rose hips.
Water	When required	

Feeding

The best method of feeding is to use a fertiliser designed for roses such as 'TopRose', however, any general purpose fertiliser will do the job.

Feed as the leaves begin to form. If TopRose is being used, sprinkle about 4oz per square metre, lightly working it into the soil surface. A couple of handfuls of Growmore or similar will help if no rose feed is available.

June : Feed as in March, but use half the amount. Water after feeding if the weather is dry.

July: Do not feed after July - this will only encourage leaf growth at the expense of flowers.

Dead-Heading

Dead-heading is simply removing dead flowers at regular intervals. This encourages the growth of more flowers later in the season. The flowers can be pinched off with your fingers or a pair of secateurs.

With floribundas, where several flowers are produced on each stem, cut the stem, just below the truss of flowers.

Weeding and Mulching

Roses have long tap roots to anchor them into the soil, but they gain most of their food from their near-surface root system. It's therefore important to keep the weeds under control. Mulching with well-rotted organic material does this job well and also provides a slow feed of nutrients.The best time to mulch is late Spring - the soil is moist but at the same time is warming up. If weeding is required, don't dig around the plants - weed by hand or on the surface with a hoe.

REFERENCES

Allen, R. C. 1948 . *Roses for Every Garden*. New York M. Barrows and Company Book Club (BCE/ BOMC) Hard Cover.

Chattopadhyay. S.K. (ed). 2007. *Commercial Floriculture*. New Delhi, Gene-Tech Books.

Jeremy Pertwee. 2000. *Production and Marketing of Roses*. Elsevier November.

21

Lotus

The Lotus (Nelumbo nucifera) symbolises purity, beauty, majesty, grace, fertility, wealth, richness, knowledge and serenity. The Pink Lotus is the National Flower of India. Nelumbo nucifera is known by a number of common names, including Sacred lotus, Indian lotus and Sacred water-lily.

Lotuses are found in white and pink colors in general and they grow in shallow and murky waters. Lotus flowers enjoy warm sunlight and are intolerant to cold weather. This is why, the Lotus is not seen blossoming in the winter. The floating leaves and Lotus flowers have long stems, which contain air spaces to maintain the buoyancy. The Lotus is native to Asia and flourishes in a wide range of climates from India to China.

The Lotus plant is an aquatic perennial, native to southern Asia and Australia, and most commonly cultivated in water gardens. The plant has its roots firmly in the mud and sends out long stems to which their leaves are attached. The leaves are sometimes, and Lotus flowers always, raised above the water surface. The beautiful and fragrant Lotus flower opens in the morning and petals fall in the afternoon.

Lotus is native to the Midle East, Asia, New Guinea and Australia. It is the most commonly featured flower in South Asian mythology and has featured in many South Asian religions through the ages. It has been cultivated since early times, for religious and ornamental purposes. In India it is commonly grown in ponds and tanks for its elegant, sweet-smelling flowers.

Lotus flowers have been used throughout history in South Asia and have been featured in Buddhist and Hindu art, architecture and literature. It was even a symbolically important plant before the religions at the time of the the Indus Valley civilisation.

The flowers became symbolic of immortality and resurrection because people observed that they would grow from the bottom of dried up pools after the monsoon rains.

Despite its early use, it was Buddhism which first brought the lotus symbol to widespread use. Lotus medallions are prominent on the Buddhist places of worship at Sanchi in Madhaya Pradesh and Amaravati in Andhra Pradesh dating from the 2nd century BC to the 2nd century AD.

As Buddhism spread from India to Central Asia and China in the first few centuries AD, lotus flowers were used to represent Buddha. They featured on rosettes, scrolls, motifs and iconography.

The giant leaves of lotus plants were used as plates in ancient India, and its seeds and roots are still considered a delicacy. 11th and 12th century texts noted lotus dishes and feasts in which lotus leaves were consumed.

Figure 1. Lotus

The lotus became a common feature woven into South Asia's culture. This continued with the advent of Islam in the 12th century AD. Lotus flowers had ancient connections with Persian culture, so they were already popular motifs on Islamic carpets, textiles and architecture. They feature in intricate patterns on perforated screens, tiles and ceramics.

The lotus is one of the world's most celebrated flowers. From ancient times to the present day, it has featured in folklore, religion and the arts in one form or another. It is a symbol of eternity, plenty and good fortune. The flowers are widely used for ornament and as offerings particularly in Buddhism and Hinduism.

The lotus has great significance in the spiritual life of South Asia. It is a symbol of purity and untarnished enlightenment amid ignorance. Particularly sacred in Hinduism and Buddhism, it represents the concept of primordial birth from the cosmic waters of creation. It is frequently mentioned in the ancient Sanskrit scriptures, where it has many names including padma (pink lotus), kamala (red lotus), pundarika (white lotus) and utpala (blue lotus).

Lotus flowers are also associated with the sun, since they open at dawn and close at dusk. The Rig Veda in around 1500 BC mentions white and blue lotuses and the Atharva Veda compares the human heart with the flower. Twin youths who are gods of dawn, known as the Ashvins, wear blue lotus garlands.

Vishnu is represented with a lotus in his navel because mythology states that a golden lotus emerged from his navel as he lay upon the primordial waters and Brahma sat upon it. A lotus also emerged from Vishnu's forehead and the lotus-hued goddess Lakshmi sprang from it. She is therefore also called Padma and iconography depicts her seated on a fully blooming pink lotus, holding a lotus and being purified by elephants. Vishnu also always carries a lotus in his hand and almost all the gods of the Hindu pantheon are seated on lotuses or carry the flower.These lotus seats, also known as a kamalasana, and lotus pedestals symolise purity in both Buddhism and Hinduism. In Buddhist mythology lotus flowers are thought to have sprang up with each step the Buddha took. The Buddhas, Bodhisattvas and Taras are often depicted holding a lotus flower or seated upon one.

The lotus embodies divine birth and is also likened to a divine womb. It is part of sexual symbolism in Vajrayana Buddhism where the union of the vajra (thunderbolt) which is hard and penetrative and the soft and open lotus are conceived as the divine embrace of compassion and wisdom.

In yoga, one of the main postures for meditation is known as the padmasana or lotus position. The head is held high and the body adopts a cross-legged seated position. This symbolises reaching above towards pure knowledge while being rooted in the material world of experience.

BOTANICAL FEATURES

The lotus *Nelumbo,* is classified as the sole member of its botanical family, the *Nelumbonaceae*. There are two species of lotus :

The lotus is a perennial plant growing from a thick rhizome. Its rounded leaves can reach up to 50 cm in diameter. The first leaves that appear, few in number, are flat and float on the surface. They are followed by thicker, funnel-shaped leaves with slightly wavy edges that may stand from 50 cm to 2 meters above the water. The leaves are coated with a film, upon which water forms magnificent, glittering droplets. The flower stalk rises above the leaves, ending in large, sweet-smelling, white or pink blooms, appearing one at a time. Each flower lasts from 2 to 5 days and darkens with age. Ranging in diameter from 15 to 25 cm, lotus flowers are termed single when they have fewer than 25 petals, semi-double if they have 25 to 50 petals, and double if they have more. After blooming, the petals fall, leaving a cone-shaped seed head that resembles the rose of a watering can. Each of its 15 to 20 openings contains a fruit.

CULTIVATION

Lotus are generally grown from rhizomes planted in late May or early June. If a lotus is to be grown in a pond or pool, it should be planted in shallow water (15 to 50 cm below the surface), directly in the bottom or in a container 30 cm tall and 50 cm across. The water temperature must be at least 18°C during the growing season. The soil must be quite rich, i.e. one part loam, one part leaf mold, compost or well-aged manure and one part clay soil. Slow-dissolving fertiliser, such as bone meal, may also be added.

The lotus prefers a very sunny location. Cooler temperatures may cause the plant to go into dormancy, and it may be difficult to coax it to rebloom. For better growth and flowering, slow-release fertiliser tablets may be used once a month, and the first flowers should be removed as soon as they fade.

The lotus is normally propagated by rhizome division, with two or three eyes being kept on each section. It can also be propagated from seed. Lotus seeds have a highly effective dormancy mechanism and require a number of specific steps to encourage them to germinate. The surface of the seeds must first be scarified to break the outer coating. Then they should be placed in a container, a drinking glass for instance, filled with a few centimeters of water, which must be changed daily. Each plant should be pricked out once two leaves emerge, along with rootlets. Plants started from seed in March may flower the first year, in August and September, while rhizomes planted in containers will flower in July.

In our climate, it is best to lift the rhizomes in the fall. They should be stored in their original soil, covered with 5 to 20 cm of water and kept cool (5° to 10°C), safe from any risk of freezing. Plants grown in the bottom of pools or ponds should be protected with sheets of plywood and polystyrene to protect the rhizomes from freezing. Containers may also be moved to a deeper part of the pond where the water does not freeze. Plants should be divided every three or four years.

Tips for Growing Lotus

This beautiful huge water-lily is not for the faint-hearted! Lotus culture is quite demanding, but it can be grown outdoors in the warmer southwestern parts of the British Isles. Otherwise, if you want to grow this plant you will need a heated greenhouse with a large pool or tub.

Hardiness

Young plants will need to be kept in a tank or tub indoors. Once they have at least five leaves, the plants can be grown in a pond outdoors, provided the roots are at least 30cm below the lowest level of water, and the pond is covered with at least 3cm of wooden boarding overwinter. Plants will not grow outdoors in parts of the UK which have prolonged or very cold frosts. Under glass the minimum water temperature must be 10°C.

Propagation

By seed. The large seeds should have a portion of the hard outer coat carefully chipped away and sown individually in pots containing ordinary potting compost. Cover the top of the pots with clean aquarium gravel so that the seed and compost will not float away.

Submerge the pots in an aquarium filled with clean rainwater containing a heater (the type sold for tropical fishtanks) which is set to give a water temperature of 15-20°C. The waterlevel should be a few centimetres above the top of the pots. Germination is usually very rapid and the young plants will grow quickly.

Keep the young plants indoors in their tank of hot water and gradually lower the water temperature to 10°C. They will need to be kept indoors for their first winter. Move them to larger pots as they grow, and plunge these into a big tub.

To grow outside, when the plants have five leaves or more, remove the roots from their pot and put in a waterlily basket filled with a mixture of two parts good soil and one part well-rotted garden compost. Cover the top of the pot with stones to stop the compost from floating away. Plunge the pot in a pond outside during the summer, when all danger of frosts has passed.

PRODUCTION AND TRADE

Lotus flowers are cultivated in ponds in South Asia as a food and for their flowers which are used for ornamental and religious offerings.

The plants are propagated by their seeds, or more commonly by their rhizomes. The rhizomes are cut into small pieces and are planted to allow a small portion to stick above the soil at the bottom of the pond. The rhizomes grow and shoot upwards to form new plants. They flower during the hot and rainy season. Flowers are cut for ornamental purposes. The edible rhizomes are ready for harvest in October.

Uses

All parts of the sacred lotus are edible. The leaves are used for serving or wrapping around other food before it is cooked. The seeds can be ground to obtain an almond-flavored flour. The stalks are often sliced and added to salads. The rhizome is eaten raw or cooked. The plant is also used for medicinal purposes. And the seed head, fresh or dried, is often used in flower arrangements.

The use of lotus in traditional Asian medicine dates back many centuries and continues largely unchanged even today. One of the most remarkable aspects of the lotus plant is that no fewer than seven different parts of the plant can be used medicinally, often for different purposes. Many of these uses are supported by scientific research.

Due to its astringent qualities, lotus has been widely used in the treatment of diarrhoea, dysentery and piles. Many traditional ancient medical texts also report its use for skin conditions, notably ringworm, and leprosy, sexually transmitted diseases such as gonorrhoea and syphilis as well as for lowering fevers, fighting fungal infections and supporting a weak heart. The milky latex found in the stems, leaves and flowers is used to fight bacterial infections.

Roots and rhizomes have been used for treating smallpox, throat conditions, loss of skin pigmentation, coughs, diarrhoea and dysentery. One preparation involves mixing boiled rhizomes with sesame oil and rubbing it on the head to cool all parts of the head including the eyes. Leaves and stems have been prepared in a variety of ways to treat piles, leprosy, parasites and vomiting. Various parts of the flower including the petals have treated diarrhoea, cholera, fever, liver conditions, bronchitis, skin eruptions, snake bites and scorpion stings. To treat coughs, syrup is made using the dried flowers. Fruits and seeds have been used to soothe inflamed mucous membranes, lower fever and get rid of bad breath. Some sources state that the seeds, taken orally with a rice wash for 7 days, can increase female fertility.

Lotus is not commonly used in western medicine. Recent scientific research has been conducted on the chemical and medicinal properties of various parts of the lotus plant. Results support its use in traditional medicine.

Scientific research on lotus provides evidence in support of many of its traditional medicinal uses to treat diarrhoea, fungal infections, fevers and conditions skin. Alcohol extracts of the rhizomes have also been shown to display as many as seven different kinds of therapeutic activity. The most notable is its anti-bacterial activity and hence use for the treatment of sexually transmitted diseases such as gonorrhoea and syphilis. Rheumatoid arthritis is also reported to re. ond well to treatment with lotus, as do certain kinds of diabetes.

REFERENCES

Desai. R.G. . 2004. *Economics of Floriculture* New Delhi, Himalaya.

John M. Dole, 1999. *Floriculture: Principles and Species*. Prentice Hall

Supriya Kumar Bhattacharjee. 2006. *Vistas in Floriculture*. Jaipur, Pointer Pub.

Watson L. and Dallwitz M.J. (1992). *The families of flowering plants: descriptions, illustrations, identification, information retrieval.* Version: 3 May 2006.

22

Gladiolus

Gladiolus (from Latin, the diminutive of gladius, a sword) is a genus of perennial bulbous flowering plants in the iris family (Iridaceae). Sometimes called the sword lily, the most widely-used English common name for these plants is simply gladiolus (plural gladioli, gladioluses or sometimes gladiolas). The genus is distributed in Mediterranean Europe, Asia, Tropical Africa and South Africa. However, the center of diversity of the genus is located in the Cape Floristic Region, where most species were discovered. As a matter of fact, 163 out of the 250 species of Gladiolus are from Southern africa. The genera Oenostachys, Homoglossum, Anomalesia and Acidanthera, traditionally considered independent entities, currently are included in Gladiolus.

The genus Gladiolus contains about 260 species, of which 250 are native to sub-Saharan Africa, mostly South Africa. About 10 species are native to Eurasia. There are 160 species of Gladiolus endemic in southern Africa and 76 in tropical Africa. The species vary from very small to the spectacular giant flower spikes in commerce. These attractive, perennial herbs are semihardy in temperate climates. They grow from rounded, symmetrical corms, that are enveloped in several layers of brownish, fibrous tunics. Their stems are generally unbranched, producing 1 to 9 narrow, sword-shaped, longitudinal grooved leaves, enclosed in a sheath. The lowest leaf is shortened to a cataphyll. The leaf blades can be plane or cruciform in cross section.

The fragrant flower spikes are large and one-sided, with secund, bisexual flowers, each subtended by 2 leathery, green bracts. The sepals and the petals are almost identical in appearance, and are termed tepals. They are united at their base into a tube-shaped structure. The dorsal tepal is the largest, arching over the three stamens. The outer three tepals are narrower. The perianth is funnel-shaped, with the stamens attached to its base. The style has three filiform, spoon-shaped branches, each expanding towards the apex.

The ovary is 3-locular with oblong or globose capsules, containing many, winged brown, longitudinally dehiscent seeds. In their center must be noticeable the specific pellet like structure which is the real seed without the fine coat. In some seeds this structure is wrinkled and with black color. These seeds are unable to germinate.These flowers are variously colored, pink to reddish or light purple with white, contrasting markings, or white to cream or orange to red.

Figure 1. Gladiolus

The South African species were originally pollinated by long-tongued anthrophorine bees, but some changes in the pollination system have occurred, allowing pollination by sunbirds, noctuid and sphingid moths, long-tongued flies and several others. In the temperate zones of Europe many of the hybrid large flowering sorts of gladiolas can be pollinated by small well-known wasps. Actually, They are not very good pollinators because of the large flowers of the plants and the small size of the wasps. Another insect in this zone which can try some of the nectar of the gladioli is the best-known European Hawk-moth Macroglossum stellatarum which usually pollinates many famous garden flowers like Petunia, Zinnia, Dianthus and others.

Gladioli have been extensively hybridized and a wide range of ornamental flower colours are available from the many varieties. The main hybrid groups have been obtained by crossing between four or five species, followed by selection: Grandiflorus, Primulines and Nanus. They make very good cut flowers. However, due to their height, the cultivated forms frequently tend to fall over in the wind if left on the plant.

CULTIVATION

In temperate zones, the corms of most species and hybrids should be lifted in autumn and stored over winter in a frost-free place, then replanted in spring. Some species from Europe and high altitudes in Africa, as well as the small 'Nanus' hybrids, are much hardier (to at least -15°F/-26°C) and can be left in the ground in regions with sufficiently dry winters. Plants are propagated either from small cormlets produced as offsets by the parent corms, or from seed; in either case, they take several years to get to flowering size.

The gladiolus is an easy-to-grow flower, especially valued for use in floral arrangements. Gladioli produce tall spikes of large blossoms, in a rainbow of colors. Only clear, true blue is missing; white, pink, red, purple, yellow, orange, salmon, and even green gladioli are available, along with many bi-colors. "Glads" grow from corms (bulb-like structures) that are not winter-hardy in Minnesota. They must either be dug in September and stored until planting time the following May, or replaced annually. Some gladiolus experts recommend treating them as annuals because you are more likely to get large, healthy blooms each year that way, and you don't have to fuss with storing them.

Soil & Climate Soil

Gladiolus produces the best spikes when grown in deep, well-drained sandy loam soil. A heavy clay soil, with poor drainage, is unfit as the gladiolus root system is easily damaged by excessive soil moisture. However, mixing of 5-8 cm thick layer of river sand along with the manure will make the clay soil cultivable. For best growth they require slightly acidic soil of about pH 5.5 to 6.5 where most of the nutrients become available to the plants.

Climate

Gladiolus prefers mild climate and sunny situation for their proper growth and flowering. Temperature is considered as a major factor influencing the number of days taken for flowering. Optimum growth of O O gladioli occurs at temperatures between 10 and 25 C where the night temperature is not above 16 C. O However, it can tolerate temperature upto 40 C only if the relative humidity is high and soil moisture levels are optimum. Light levels affect initiation of flower. The period of flower initiation commences when the third leaf becomes visible and ends when the sixth and seventh leaf appears. Quality of flower spikes and yield is better in a long day condition than short days. Minimum illumination of 8 hours/day is essential for most of the gladiolus varieties.

Planting Layout

If you are growing glads primarily for cut flowers, you may want to plant them in rows, as you would vegetables. It is easier to prepare the area, stake and tend the plants, and harvest flower stalks when they grow in neat rows. Glads can also be used to provide color in annual beds and borders, though they are very stiff and upright and a bit awkward to work with. Plant gladioli in groups of seven or more corms of the same cultivar (cultivated variety) for best effect.

Site

Choose a location in full sunlight. Although newly purchased corms are ready to bloom and should flower even in the shade, flowers will be larger and brighter and stalks will be sturdier when they're planted in sun. The glads will also be able to store more energy for the following year's bloom, which is critical if you plan to re-use your corms. Well-drained soil is essential for successful gladiolus growing. If your soil is heavy or tends to be wet, create raised beds for your glads (and most other annuals, perennials, and bulbs). Whether or not you garden in raised beds, loosen the soil to a depth of ten or 12 inches. Fertilize, if necessary, according to recommendations based on a soil test.

Land Preparation

The soil should be thoroughly ploughed 30-40 cm deep and exposed to sun for at least 25 days. After removing the weeds, the field should be reploughed and levelled. If green manuring crop is taken it should be turned down into the soil at least two months before planting of corms. Animal manure is not recommended, as it is a source of pathogens. Ridges and furrows are prepared as per the spacing recommended.

Size of Corms

The size of the corms has a positive correlation with the height of the spike and number of florets per spike. Medium-sized corms of 4-5 cm diameter each are selected for planting to obtain quality blooms and uniform flowering. The bigger corms called 'Jumbo', 6-8 cm diameter, are good for getting top quality spikes for exhibition purposes. Smaller corms, of 2-5 cm diameter each can also be planted as flowering stock but corms and cormels having diameter less than 2.5 cm are mainly used as a planting stock in the nursery for development of bigger corms for next year planting.

Planting

Start planting in mid-May, then again every two weeks through mid-June. This schedule will keep the flowers coming form July through August. You could also choose early, mid-season, and late cultivars, plant them all in May, and still enjoy continuous bloom for much of the summer. The final strategy to extend bloom time would be to plant different sized corms. Larger corms bloom somewhat earlier than smaller corms of the same variety.

Corms smaller than ¾ inch in diameter may not produce flowers. To ensure large-sized blooms, plant corms that are 1¼ inch or larger in diameter. Choose corms that are relatively tall and plump, shaped like a chocolate kiss, rather than wide and flat. Thick corms produce good quality flowers.

Plant corms with the pointed side up, about four times as deep as their diameter. Measure the distance to the bottom of the planting hole, then plant the corm. (A one-inch corm would be planted four inches deep, a 1¼-inch corm, five inches deep, and so on.) Space the corms six to eight inches apart.

Label your corms and put in stakes or other support structures when you plant them. Gladiolus flower spikes blow over easily in the wind. If you stake them first thing, you can avoid the likelihood of damaging their roots with the stakes.

Time of Planting

The best time for planting flower size corms is mid-September to mid-October. For getting a succession of flowering spikes from November to May, corms are planted from early September to the end of November, at fortnightly intervals. In hilly regions, the planting time is from March to May so as to get the spikes from June to October.

Method of Planting

Normally, corms are planted 8-15 cm deep at a spacing of 15-30 cm x 30-40 cm, depending upon the size of corms. A plant population of 15/sq.m or 1,50,000/ha is ideal for quality production of flower spikes and corms. The following are the corm size/density guidelines-The soil should have adequate moisture before planting so that no watering is required till sprouting. Each cultivar has a fairly definite number of days to bloom, ranging from 60-69 days, 70-74 days, 80-84 days, 91-99 days and 100 days or more for very early, early, early mid-season, mid-season, late and very late cultivars, respectively.

Weeding and Hoeing

The field is kept clean by regular weeding and hoeing. When the area is small, weeding and hoeing is done manually. First weeding is completed within three weeks after sowing while the second weeding is done before application of fertilizer as top dressing.

Earthing Up

Earthing up is practiced when the soil is heavy and deep planting is not possible. It is done when the plants are 20-30 cm in height. The tall plants are staked. Staking is done after the emergence of spikes but before opening of florets. Plants are loosely fastened at three places with the help of jute cord.

Mulching

Mulching with dry grass helps to conserve the moisture and suppress the weed growth. However, this practice should be discouraged in fields infested with termites.

Irrigation

Sufficient water is required for good growth of gladiolus. Number of irrigation depends upon the climate and type of soil. Usually, 1-2 irrigation a week are sufficient. Irrigation is provided immediately after the application of fertilizers and during the emergence of buds. Watering should be stopped at the time of ripening of corms.

Manuring & Fertilization

Application of fertilizer leads to the formation of healthy and large sized corms, which give large-sized flowers on long spikes during the succeeding year. FYM should only be applied for a preceding crop and never before planting gladioli. Application of fertilizer use should be based on soil analysis. Normally 120 kg N, 150kg P_2O_5 and 150 Kg K_2O per hectare is recommended, of which 60 kg N and entire dose of P_2O_5 and K_2O is applied as basal dose. The balance N is given in two split doses, one as foliar spray at four leaf stage while the second as soil application at the bud stage. Application of growth regulators mainly GA_3, IAA and NAA as soil drench has beneficial effects on the growth and flowering of plants and induces early flowering.

Summer Care

Apply a layer of mulch such as straw, grass clippings, or pine needles to help keep weeds down. Pull or hoe any weeds that come up. Mulch will also help conserve moisture in the soil, cutting down on surface evaporation. Adequate rainfall or watering is still needed for best quality blooms, so be sure your glads receive an inch or water each week, if possible.

Cutting for Bouquets

Bring a sharp knife or florist's shears and a tall bucket of lukewarm water to the garden with you. Cut the flower spikes first thing in the morning or at night, not during the heat of day. Cut spikes with only one, two, or possibly three flowers open; the rest will open in order, up the spike. Allow at least four leaves to remain on the plant if you wish to re-use the corms. Cut diagonally through the stalk and place it in the lukewarm water immediately. Once you've collected all the glads you want cut, put the bucket in a cool, dark place for a few hours so the blooms "harden off". Use floral preservative in the vase water before arranging the glads. As lower flowers fade, nip them off. Cut about an inch of stem off the bottom of each spike every few days.

Storage

Dig gladiolus corms once the foliage has been killed by frost. Shake off excess soil and sort the corms by cultivar. Cut the stem off just above each corm. "Cure" corms for about three weeks in a warm, dry, airy place. At this point, the corms you planted in spring will easily break off the bottoms of the new corms that developed over the summer growing season. Discard the old, spent corms and save the new ones. Leave their husks intact, and treat them for insects if they appear infested. (See "pests" following.)

Place the new corms in paper bags, cloth sacks, or nylon pantyhose legs. Store them in a well-ventilated place that's dark, dry, and cool. Ideal storage temperatures range from 35° to 45° F; the cooler the better, as long as they are not allowed to freeze.

When you dig the corms, you'll notice a number of miniature corms attached to the main one. These are called cormels. They should be able to bloom in two or three years if you save them and replant them each spring. Save the largest ones, at least ½ inch in diameter. Plan to plant them about 1½ to 2 inches deep.

PESTS AND DISEASES

Glads are susceptible to a number of diseases and are prey to insects, as well. To minimize the chance of disease or insect problems, always start with sound corms. Toss any that look odd or feel soft or crumbly. Practice "crop rotation" if possible, planting glads in different locations from one year to the next. If plants are yellow or stunted assume the worst—virus infection, for which there is no cure—and destroy them.

If the leaves appear streaky, or if flowers fail to open or are misshapen or streaked and discolored, the problem is probably thrips. Thrips are tiny insects that overwinter on stored corms. They use rasping mouthparts to feed on gladiolus foliage and flowers, often while the flowers are still in the bud. Spray the plants when you first see damage, using acephate (Orthene and others) or carbaryl (Sevin). Repeat according to label directions if damage continues.

Most of the better glad growers are now treating their bulbs for thrips before sale, so it is not necessary to dip or soak bef planting. But to be on the safe side use a chemical to insure that no thrips or bacterial diseases are going to be carried into the patch from new bulbs.

When the plants are about ten inches high, begin spraying or dusting with a good insecticide, even though you have had no trouble with thrips. These little pests when mature are only about a sixteenth of an inch long. They suck the juices from the buds, which look as though they had been burned. The florets do not open properly, or if they do, they have a sickly appearance.

Gladiolus Pests Thrips

This is one of the most serious pests of gladiolus. Yellow coloured nymphs and black adults damage leaves and spikes. Affected leaves and spikes develop silver streaks, turn brown, get deformed and dry if damage is severe. Attack on young plants reduces flower production. The pest also attacks corms under storage. Infected corms become sticky, shrivel and produce weak plants when planted.

Control: Spraying of Methyl Demeton 25EC or Dimethoate 30EC @2ml/litre of water at 10 days O interval provides significant control of thrips. Storing of infested corms at 2 C for 6 weeks and later O treating them with hot water at 46 C completely kills the thrips on gladiolus corms.

Cut Worms

The pest attacks mainly the newly planted gladiolus plants. Female moth lays eggs near ground on plant parts. Hatched larvae feed during nights on emerging shoots. Grown up larvae, which are clay coloured, cut the plants at ground level. Plants are vulnerable to attack upto 3rd leaf stage. They also damage underground corms and developing spikes.

Control: Ploughing during summer exposes the pupae to predators. Poison bait consisting of Carbaryl or Malathion @ 0.1%, wheat bran and molasses scattered in the field effectively controls the larvae. Sprays of Methyl Parathion @0.05% or Quinalphos @0.05% provides protection to foliage from cutworms

Leaf Eating Caterpillar

This caterpillar mainly damages foliage of gladiolus plants. Female moth lays ash coloured eggs in groups on lower side of leaves. The young larvae feed on lower surface of leaves by scraping while greenish-brown mature larvae feed voraciously during nights on these leaves.

Control: Collection and destruction of egg masses and leaves infested with young larvae reduce pest build up. Setting up of light traps attracts adult moths and helps in monitoring pest population. Deep ploughing in summer exposes pupae to predators. Spraying of Quinalphos @0.05% or Carbaryl @0.1% or Chlorpyriphos @0.05% gives protection to foliage from the leaf-eating caterpillar.

Mites

Attack by the mite begins at an early stage when plants are young. Green coloured nymphs with lateral specks and reddish-brown coloured adults colonize on leaves and suck the sap causing discolouration and wilting. The affected leaves fall off.

Control: Methyl Parathion 0.05, Fluvalinate 0.012% and Diazinon at 0.4% offer significant control of mites.

Mealy Bug

Damage by mealy bug begins in the field on underground corms during dry conditions and carries on to the storage. Nymphs and adults damage corms by sucking the sap causing shriveling and drying of affected corms.

Control: Prompt collection and destruction of infested parts reduces spread of the pest. Crawling of ants on plants is the sign of beginning of mealybug infestation. Spraying should be taken up at this stage. Sprays of Methyl Parathion 0.04% or Dimethoate 0.04% or Acephate 0.1 % at 15 days interval effectively contains mealybug infestation

Root-knot Nematode

The symptoms of the attack are in the form of stunted growth, yellowing of leaves, heavy galling on roots.

Control: Use nematode-free planting material is the best method to avoid the nematode infestation. Hot water treatment of corms at 57.8°C for 30 minutes helps to control the nematode population. Intercropping or crop rotation with marigold reduces the severity of attack in the field. Application of 2 Carbofuran/Phorate @ 1g a.i./m is found effective.

Botrytis Soft Rot

The disease occurs in the cold storage when the temperature stays below 13 -15 C. The disease is characterized by the formation of white-molded, soft, spongy corms. At early stage, the infection may be confined to surface only, but later on, it travels to the core as well. At high temperature, corms seal off infection, escaping further rotting. In the field, infection is characterized by circular brown leaf spots. The smallest spots are visible only on the upper surface of leaf.

Control: Leaf infection cab be controlled by spraying the plants with 0.25% Dithane M-45, twice a week and by dusting healthy corms with Dithane M-45, during storage.

Fusarium Corm Rot

The fungus is carried in corms /cormels, and the soil. Latent infection in corms may develop into active rotting of corms/cormels during storage and also during plant growth. Extra losses occur when plant growth is checked by poor soil aeration, flooding or hot weather. Addition of excess nitrogen in the form of cow dung manure results in additional casualties. The disease is characterized by rot of corms, stunted plant growth, late blooming or blind plants, greener and smaller buds and poor opening. The most serious loss is the reduction in number and quality of flower spikes due to the infection.

Control: Fusarium corm rot is difficult to prevent but can be controlled to some extent by growing disease tolerant cultivars and by various control measures. Hot water treatment of corm /cormels at a 0 temperature of 50 C for 30 minutes before planting, use of clean soil or fumigated soil, and treatment of corms /cormels with fungicide prevent/ control the disease. The most effective treatment is soaking the corms in Benlate/ Bavistin 0.02% solution for half an hour before planting and soil application of Furadan/ Thimet 10G @ 3g/square metre at the time of planting is very helpful in preventing the incidence of this dreaded disease.

Penicillium Rot

This is commonly known as green mold. The disease occurs when the corms are stored in heaps after having been dug in the cold weather. Large reddish brown lesions on the sides of the corm characterize the disease. Sometimes lesions may be covered with bluish green masses of fungal spores. Hot humid conditions favour the disease.

Control: The corms during storage should be examined periodically and disease corms should be sorted out and discarded. Storing of the healthy corms may prevent the occurrence of the disease. Dusting of corms with Dithane M-45 is helpful in preventing this disease during storage.

HARVESTING

Gladiolus spikes are ready for harvesting when the first bud shows the colour of the variety. Normally, th th early varieties start producing spikes from 75 day while the late season varieties from 85-90 day after planting Spikes should be cut in the tight bud stage, with one or two pairs of leaves and 1-3 floral buds showing colour. The plant should be left with 3-5 leaves on the stem for the development of new corms and cormels. If the spikes are to be used as fresh flowers, then they should be cut with 3-4 open florets, but without any sign of withering. The spikes are cut with a sharp knife either early in the morning or in the evening. Slant cut should be given at the base of the harvested spikes and immediately placed in a bucket with cold water.

Lifting of Corms and Cormels

Corms planted during September-October are ready for lifting during March-April in the North-Indian plains and upto August in hills. Corms are matured when 25% cormels have become brown which generally take 30 to 45 days from flowering when the leaves also start yellowing. Plenty of moisture, followed by a dry period, before lifting ensure the formation of large corms. After flowering, when the leaves start turning yellow, plants are twisted down to ground level for allowing the corms to mature. No irrigation should be given thereafter.

Corms and cormels should be dug out with the help of a spade. Soil should be dug deep in order to take out all the cormels. The corms are checked for any disease infection and the affected corms are discarded.

Curing

Curing is one of the essential post harvest operations for successful storage of corms. After lifting and removing the adhering soil, the corms and cormels of each cultivar are kept in trays in a shady but well ventilated place for about a fortnight. For curing, the layers of corms should not exceed three, which may be cured for five weeks at 21°C.

Cleaning, Grading and Storage

After the corms are fully cured, these are cleaned and diseased ones discarded. The old withered corms are taken out and cleaned. Treating the corms with 0.2% Captan 15 days before storage or dusting with 5% Cythione dust and Dithane M-45 protects them from insects, pests and diseases during storage. After cleaning, the corms and cormels are graded in different grade-sizes.

The corms are stored in perforated trays in a well-ventilated cool and dark room with temperatures not exceeding 27 C. Being smaller in size, the cormels are stored in plastic trays having fine perforations. It is advisable to keep on turning corms and cormels periodically, for preventing their rotting due to poor aeration. The corms are periodically checked during storage and the decaying ones are removed.

The cormels of exotic cultivar vary considerably in respect of hardiness, depending upon their size. The small sized cormels (<0.5cm dia) are fairly hardy and may be stored at room temperature, without decay. However, the large ones (>0.5cm dia.) of exotic cultivar require low temperature during storage and should be kept in cold storage. To avoid chances of mixing, the cormels are packed in hessian cloth bags before putting them in perforated trays for keeping them in the cold storage. Like corms, the cormels should also be taken out of the cold storage in the first week of October and kept at room temperature for a week before planting them in the ground.

Yield

The yield of flower spikes and corms in gladiolus depends on variety, corm size, planting density and management practices. Gladiolus planted at a spacing of 30 x 20 cm yields approximately 1,50,000 marketable spikes per hectare. Additional income can be obtained from the sale of about 3.5 lakh corms.

HOW TO SUCCEED WITH GLADIOLUS

How many times have you admired a beautiful spike of gladiolus, or an arrangement of glads, and wished that you too could grow similar flowers? It isn't as difficult as one might imagine.

Gladiolus can be grown on any soil that will grow weeds, from a heavy clay to a light sand. Among the few requirements are lots of sunshine and plenty of water. One of our most widely quoted authorities says that the best "fertilizer" for glads is water. However, they as well as the weeds do much better with a fertile soil. A slightly acid soil seems to be to their special liking. It is important to select the best varieties and the best corms available. Your climate determines the varieties which will give you best results.

Although some of the very best spikes may be grown in Canada and Alaska, it would be asking too much to expect some varieties to mature in the short growing season there. Glads go from planting to flowering in 65 to 100 days, depending on variety and locality. Begin planting in the spring as soon as the trees begin to leaf out. One rule is "plant when the oak leaves are the size of a gopher's ear." For season-long bloom, you may plant at two-week intervals, or you may use various size corms.

Depth of planting depends on the type of soil in which corms are planted. Sandy soils require greater depth than heavier soils. It is well to plant the larger or number one size at a depth of from six to eight inches for light soils, and proportionally less in heavier soils.

A good plan is to plant in a trench, covering the corm with only an inch or two of soil. Then as cultivation continues, the soil will be covering the plants as they emerge. This practice also reduces the amount of weeding to be done. Be sure to keep the patch weed free. This is essential, not only because weeds take as much food from the soil as do glads, but also because thrips, those nasty little insects, seem to attack in fields or patches where there are weeds, more than in patches kept clean.

Several weed control chemicals have been on the market for a time long enough to determine which is best suited to your particular problem. For the home garden it is good to be wary of any without first determining whether there might be damage done to the glads.

REFERENCES

Desai. R.G. . 2004. *Economics of Floriculture* New Delhi, Himalaya.

Dole, John M. and Harold F. Wilkins. 1999. *Floriculture: principles and species.*

Everett, T. H. (ed.) 1964 . *New Illustrated Encylopedia of Gardening.* New York Greystone Press.

Gast, K.L.B., et al., 1994. *Cold Storage for specialty cut flowers and plant material* (MF-1174), Cooperative Extension Service, Kansas State University, Manhattan. How to build your own cooler.

Bibliography

Billy, Simon. *Horticultural Farm Manager*. The Kerr Center for Sustainable Agriculture, Poteau, OK. Personal communication. May 1997

Bordelon, B. 2002. *Grape Cultivars for the Upper Midwest*. Purdue University, Department of Horticulture and Landscape Architecture

Bostanian, N. J., M. Binns, J. Kovach, G. Racette and G Mailloux. 1999. "A predictive model for strawberry bud weevil adults in strawberry fields". *Environ. Entomol* 28: 388-406

Buck, P. A.. "Origin and Taxonomy of Broccoli" (in English). Department of Food Technology, University of California. http://www.springerlink.com/content/ert85x3082740212/fulltext.pdf. Retrieved on 2009-05-14

Carley Petersen and Annabelle Martin. "General Crop Information: Persimmon". University of Hawaii, Extension Entomology & UH-CTAHR Integrated Pest Management Program. http://www.extento.hawaii.edu/kbase/crop/crops/i_persim.htm. Retrieved on 2007-01-15

Carvalho, S. M. P., H. et.al. 2005. "Temperature affects chrysanthemum flower characteristics differently during three phases of the cultivation period". *Journal of Horticultural Science and Biotechnology* 80(2): 209-216

Chan, h.t. Jr. 1984. (Ed.). *Handbook of Tropical Foods*. New York/Basel: Marcel Dekker.

Chandler, William Henry. 1958. *Evergreen Orchards*. Philadelphia: Lea and Febiger.

Daniel Zohary and Maria Hopf, 2000. *Domestication of plants in the Old World*, third edition Oxford: University Press.

Dixon, G.R. 2007. *Vegetable brassicas and related crucifers*. Wallingford: CABI

English-Loeb, G., M. P. Pritts, J. Kovach, R. Rieckenberg and M. J. Kelly. 1999. "Compensatory ability of strawberries for bud and flower removal: Implications for managing the strawberry bud weevil". *J. Econ. Entomol*. 92: 915-921

Food and Agriculture Organization of the United Nations. 1986.*Improvement of post-harvest fresh fruits and vegetables handling*. FAO,Bangkok, Thailand.

Galán Saúco, Víctor. "Current Situation, Trends, and Future of Agronomic Research on Tropical Fruits." In Proceedings of the International Conference on Tropical Fruits, Kuala Lumpur, Malaysia, 23–26 July 1996.

Galleta, Gene J., and David G. Himelrick, (eds.) 1990. *Small Fruit Management*. Englewood Cliffs, N.J.: Prentice Hall.

Gould, W.A. 1992. *Tomato production, processing, and technology*. ThirdEd. CTI Publications, Inc, Baltimore.

Grierson, D., and A.A. Kader. 1986. Fruit ripening and Quality. Pp. 241-280. In: *The Tomato Crop* (J.G. Atherton and J. Rudich, eds.) Chapman and Hall,London.

Grigson, Jane (1978). *The Vegetable Book*. London: Penguin

Grubben, G.J.H. & Denton, O.A. 2004. *Plant Resources of Tropical Africa 2. Vegetables*. PROTA Foundation, Wageningen; Backhuys, Leiden; CTA, Wageningen

History of the Chrysanthemum. National Chrysanthemum Society USA

Huxley, A., ed. (1992). *New RHS Dictionary of Gardening*. Macmillan ISBN 0-333-47494-5

Jackson, David I., and Norman E. Looney. 1999. *Temperate and Subtropical Fruit Production*. 2d ed. Wallingford, Oxon., U.K.: CABI.

Jason Zasky. "Turning Over a New Leaf Change from 'Prune' to 'Dried Plum' Proving Fruitful". http://www.failuremag.com/arch_business_dried_plums.html. Retrieved on 2008-01-26

Kader, A.A., and L. Morris. 1976. Appearance factors other than color andtheir contribution to quality. Pp. 8-14. In: Proceedings of the Second TomatoQuality Workshop. University of California, Davis, USA

Kovach, J., R. Rieckenberg, G. English-Loeb, M. Pritts. 1999. "Oviposition patterns of the strawberry bud weevil (Coleoptera: Curculionidae) at two spatial scales and implications for management." J. Econ. Entomol. 92: 1358-1363

Mauseth, James D. 2003. *Botany: an introduction to plant biology*. Boston: Jones and Bartlett Publishers.

Molly Damon, Nancy Z. Zhang, David B. Haytowitz, Sarah L. Booth (2005). "Phylloquinone (vitamin K1) content of vegetables". *Journal of Food Composition and Analysis* (Elsevier) 18: 751–758

Nagy, Steven, and Philip E. Shaw. 1980. *Tropical and Subtropical Fruits:Composition, Properties, and Uses*. Westport, Conn.: Avi.

Popenoe, Wilson. *Manual of Tropical and Subtropical Fruits*. 1920. New York: Hafner Press.

Roach, F. A. 1985. *Cultivated Fruits of Britain*. Oxford: Blackwell.

Ryall, A.L., and W.J. Lipton. 1979. *Handling, Transportation, and Storageof Fruits and Vegetables*. AVI Publishing, Westport, Conn.

Ryugo, Kay. 1988. *Fruit Culture: Its Science and Art*. New York: Wiley.

Stevenson Whitcomb Fletcher, *Strawberry Growing*, The Macmillan Co., New York, 1917

Szalay, L., Papp, J., & Szaóbo, Z. (2000). Evaluation of frost tolerance of peach varieties in artificial freezing tests. In: Geibel, M., Fischer, M., & Fischer, C. (eds.). *Eucarpia symposium on Fruit Breeding and Genetics*. Acta Horticulturae 538. Abstract

Tall JM, Seeram NP, Zhao C, Nair MG, Meyer RA, Raja SN (2004). "Tart cherry anthocyanins suppress inflammation-induced pain behavior in rat". *Behav. Brain Res.* 153 (1): 181?"8. doi:10.1016/j.bbr.2003.11.011. PMID 15219719

Van Der Ploeg A. and E. Heuvelink. 2006. "The influence of temperature on growth and development of chrysanthemum cultivars: a review". *Journal of Horticultural Science and Biotechnology* 81(2): 174-182

Vaughan, J.G.; Geissler, C.A. (1997). *The New Oxford Book of Food Plants*. Oxford University Press

Villareal, R.L. 1980. *Tomatoes in the Tropics*. Westview Press,Boulder, Colorado.

Watson L. and Dallwitz M.J. (1992). *The families of flowering plants: descriptions, illustrations, identification, information retrieval*. Version: 3 May 2006.

Wim J. M. Koopman, Eli Guetta, Clemens C. M. van de Wiel, Ben Vosman and Ronald G. van den Berg (1998). "Phylogenetic relationships among Lactuca (Asteraceae) species and related genera based on ITS-1 DNA sequences". 1517–1530

Index

Agri-export zone 17
Agro-climatic requirements 18
Aleutian low 3
Anthrophorine bees 306
Anti-bacterial activity 304
Anticyclone 3
Apple cultivars 31
Apple cultivation 17
Apricot cultivars 35
Atharva Veda 301
Atmospheric pollution 8

Bagging 23
Biocontrol mechanisms 85
Biofungicides 85
Bisexual flowers 305
Black forest 8
Black-earths 5
Bodhisattvas 301
Boreal forest 5
Botrytis Soft Rot 313
Brassica family 165
Brown rot 81
Buddhism 300
Business proposition 11

Cabbage root fly 170
Cauliflower 165
Central leader trees 19
Chlorofluorocarbons 8
Clean-air legislation 8
Climato-vegetational zones 1
Club root 171
Coal-burning 8
Commercial apple industry 29
Companion Planting 172
Continental climate 1
Cool-weather crop 166
Crop rotation 171
Cut flowers 306
Cyanogenetic glycosides 77
Cyanogenic glycosides 36
Cyclones 2

Deciduous temperate forests 5
Deciduous trees 9
Dormant Pruning 20
Dwarfing rootstock 29

Environmental succession 5
Extreme heterozygotes 17

Fall-planted trees 41
Flea beatles 172
Fruit storage 28
Fruit thinning 43
Fusarium corm rot 313

Goldcot 40
Good sanitation practices 26
Graft-transmissible problems 35

High-density apple orchard 30
Himalayan range 18
Hinduism 300
Horticultural fleece 172

Iconography 300
Insect control 26
Intermediate latitudinal position 1

Just caterpillar 171